MATHEMATICAL PUZZLING

A. Gardiner

DOVER PUBLICATIONS, INC.
Mineola, New York

Bibliographical Note

This Dover edition, first published in 1999, is an unabridged republication (with three minor corrections) of the work originally published by Oxford University Press, 1987.

Library of Congress Cataloging-in-Publication Data

Gardiner, A. (Anthony), 1947–
   Mathematical puzzling / A. Gardiner.
      p. cm.
   Originally published: Oxford ; New York : Oxford University Press, 1987.
   ISBN 0-486-40920-1 (pbk.)
    1. Mathematical recreations. I. Title.
QA95 .G155 1999
793.7'4—dc21   99-045604

Manufactured in the United States of America
Dover Publications, Inc., 31 East 2nd Street, Mineola, N.Y. 11501

To
Tom Beldon

'Where there are problems, there is life.'
The Radiant Future, Alexander Zinoviev

Preface

'What was lacking in the usual approach, even at its best, was any sense of genuine enquiry, or any stimulus to curiosity, or appeal to the imagination. There was little feeling that one can puzzle out an approach to fresh problems without having to be given detailed instructions.'
Aspects of Secondary Education, HMSO 1979

Many of those who read this book will be interested in the puzzles and investigations for their own sake. Everyone will, I trust, find some things that are new – perhaps even refreshing. And I hope that those who teach mathematics will find the collection especially useful. The problems will challenge and stimulate able youngsters in a way which will make them think. The content, style, layout, and sequencing of the material have been chosen to meet, as far as seems possible, the needs of the individual reader, working alone. But it is not like an ordinary text, and the reader should not expect to work steadily through from beginning to end. It is a book you can begin at one age and pick up again years later.

Many of the problems and ideas can be, and have been, used with wider ability groups. Teachers will want to pick and choose (and adapt!) those problems, investigations, or sections which they judge to be suitable for use with whole classes.

The items have been formulated, and grouped together, to bring out *the mathematics behind the puzzles*. However, the word 'mathematics' here has to be interpreted somewhat more broadly than is usual. The puzzles exploit standard properties of numbers and shapes. But they also seek to develop more elusive mathematical skills, including:
- active interpretation of a problem whose statement (deliberately) fails to spell out certain natural assumptions or unnecessary details
- intelligent searching for promising starting points in unfamiliar problems
- recognizing the presence (or absence) of significant patterns within a problem or group of problems; identifying unexpected similarities between otherwise dissimilar problems; or striking differences between superficially similar problems.

The statement of a particular problem may appear unfamiliar, unclear, or ambiguous – yet, on reflection, admit of only one sensible interpretation. It may fail to advertise any obvious method of solution – yet a path may open up once you begin to chew over the connections between what you know and where you want to get to. It is all too easy to dismiss such unfamiliar problems not just as hard, but as *too* hard. They are not.* The title *Mathematical Puzzling* is meant to suggest that certain ways of thinking which we use when puzzling out this kind of problem are themselves part of 'the mathematics behind the puzzles'.

When we learn, or teach, school mathematics we usually concentrate on one or two ideas at a time and try to master them in some paradigmatic form: that is, in the context of a standard problem framework which seems to capture the essential features of the relevant idea or technique. This is all perfectly reasonable. But a balanced mathematical education must seek to do rather more. This collection may be seen as one attempt to serve some important, but commonly neglected, ends.

For those teachers who recognize the eternal tension between lowly means and higher ends, puzzles and investigations can play a dual role by providing both a stimulus for students and an invaluable diagnostic tool for teachers. A regular diet of carefully chosen puzzles and investigations – tackled individually or in pairs and discussed corporately – can help to integrate simple techniques and complex problem-solving strategies in an effective and satisfying way. The spirit of 'puzzling things out', the 'sense of genuine enquiry', the 'appeal to the imagination' – and the kind of activities which go with them – help to keep students on their toes and to remind teachers of the extent to which those higher ends are being served.

The problems challenge the able student. But they are also tractable. None of them *requires* the use of a calculator or a microcomputer, though some naturally invite the use of such aids. When, and whether, one makes use of such machines for a particular problem is a matter of taste and habit. But the *initial* exploration and analysis of the problems here should always be unaided. And whether or not one reaches for a calculator or micro at some stage to help investigate a problem, there remains the fundamental challenge of explaining, or justifying, one's eventual findings.

Most of the problems have proved their worth with various groups of children and adults over the last ten years. I am indebted to all these willing guinea-pigs. And to Steve Drywood for his sensitive and painstaking criticism of the final draft. Among the many teachers who have sought to 'educate' me during this time I should like to single out those members of the SMP 11–16 writing team with whom it was my privilege to work for so long; and Tom Beldon – to whom it gives me the greatest pleasure to dedicate this little book.

---

*To help convince potential users of this collection, Section 30 looks at four problems which have been so labelled by various groups of teachers and other adults. These problems are not meant to be easy, but they are certainly not 'too hard'! One must, however, realise that no written solution can do justice to the two stages referred to earlier: the struggle first to interpret, and then to find a way into a given problem.

Contents

How to use this book

The Problems

1. The problems are grouped into twenty-nine sections. Early sections tend to be easier than later ones, but there is no strict dependence between them.

2. The first few problems in each section are considerably easier than later problems in the same section. A few problems which seem substantially more difficult than others in the same section are marked with a *.

3. The sections are of three types, each with its own logo.

Sections 1, 4, 7, 10, and so on, are collections of problems which have a common underlying theme or 'flavour', but which may be otherwise unrelated. It may, however, still be wise to begin by tackling some of the earlier (easier) problems in such a section before going on to the later (harder) problems in the same section.

Sections 2, 5, 8, 11, and so on, explore one particular idea from a variety of different viewpoints. The later problems in one of these sections may not be strictly dependent on the earlier problems, but they do tend to build on the experience you would have gained by tackling those earlier problems.

Each of the sections 3, 6, 9, 12, and so on, is an extended investigation of a substantial piece of mathematics: the sequence of individual questions contributes little by little to the posing, and possible solution, of some central problem. In these sections you should definitely begin at the beginning (though there is no need to feel that you have to keep going right to the end).

4. You should not feel that you have to plough through all the problems in each section, one after the other. But since the problems in a section are often connected, sometimes in unexpected ways, those who do work through a whole section may well discover the kind of unexpected connections which are of the essence of mathematics.

5. A successful puzzle must appeal to the imagination, though it may not be immediately obvious how to go about solving it. However, as you begin to explore each problem in this collection it should gradually become clear how to proceed.

6. The problems provide plenty of scope for exploration and investigation. However, relatively few of the problems could be called open-ended. But behind the carefully structured façade of each section there lurk various ragged, and possibly more interesting questions. Some of these have been singled out as **Investigations** which appear at the end of the corresponding section of Problems. Others arise out of the attempt to solve some particular problem and cannot easily be taken out of that context: these **Investigations** and **Extensions** are printed as and when they arise as part of the Commentary.

The Commentary

To each group of Problems there is a corresponding section of Commentary, which contains hints and comments on most of the problems. It is up to you to use the Commentary sensibly. You should always have a good go at each problem before looking at the Commentary on that problem; but the Commentary is there to help you, so don't be afraid to use it. Even when you solve a problem entirely on your own, you may still find it worthwhile to read the Commentary afterwards.

1

Opening number

1. You probably know the Christmas song which starts like this: 'On the first day of Christmas my true love gave to me . . .'. How many gifts did my true love give me altogether during the 'Twelve Days of Christmas'?

2. Pages 6 and 19 are on the same (double) sheet of newspaper. Which other pages are on the same sheet? And how many pages does the newspaper have altogether?

3. Ever have trouble remembering your age? Never mind! Take your house number and double it. Add 5. Multiply by 50. Then add your age, the number of days in the year, and subtract 615. The last two digits are your age, the first two are your house number. Can you explain why?

4. If all the stars stand for the same number, can you complete:

$$\frac{*}{*} - \frac{*}{6} = \frac{*}{12} ?$$

5. In 12 = 3 × 4 the four successive digits 1, 2, 3, 4 occur in the right order. Can you find another equation like this one which also uses four successive digits in the right order?

6. (a) Place the numbers 1 to 6 in the six circles so that adjacent pairs, like 1 and 2, or 4 and 5, go in circles which are *not* joined by a line.

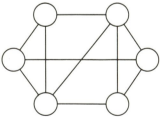

(b)   Place the numbers 1 to 8 in the eight circles so that adjacent pairs go in circles which are not joined by a line.

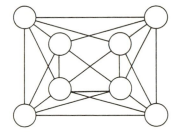

7.   You may never have noticed, but it just so happens that

$$\sqrt{(2\tfrac{2}{3})} = 2\sqrt{(\tfrac{2}{3})}$$

How many other equations are there like this one?

8.   The number 12 is equal to exactly four times the sum of its digits. So is 24.
   (a)   Can you find a whole number which is equal to exactly twice the sum of its digits? Is your answer the only possible answer?
   (b)   Can you find a whole number which is equal to exactly three times the sum of its digits? Is your number the only one?
   (c)   Which numbers other than 12 and 24 are equal to exactly four times the sum of their digits?

Investigation 1 (Do Question 8 first.)
Which numbers are exactly divisible by the sum of their digits?

---

* **Investigation 2**
Work out $2^5 \times 9^2$. What do you notice? Does any other similar arrangement of four digits have the same property?

Commentary

1.   How many 'partridges in a pear tree' does my true love give me altogether? And how many 'turtle doves' does s/he give me in all?

2.   As in all such problems, if you find it difficult to imagine the newspaper, try making a model.

3.   Choose any starting number. What effect does the sequence of instructions double it, add 5, multiply by 50, add 365, subtract 615' have on your starting number? Can you see why?

4.   What must $\frac{*}{*}$ be equal to?

6.   (a)   Think carefully where, or where *not*, to put the number 1. Which number should you think about next? Some numbers cannot be put in certain positions. Work out which and it will be easier to place the others.

7. There are lots. Try to find them all – and don't be afraid to use a little algebra. (Notice that $\sqrt{(2\frac{2}{3})}$ means 'root two *and* two thirds' that is, $\sqrt{(2 + \frac{2}{3})}$ whereas $2\sqrt{(\frac{2}{3})}$ means $2 \times \sqrt{(\frac{2}{3})}$.)

8. (a) Intelligent trial and error should find one such number. But to decide whether there are any others you will have to think a little harder. (What is the largest possible digit-sum for a number between 20 and 30? Could any of these numbers be equal to exactly twice the sum of their digits? What is the largest possible digit-sum for a number between 30 and 40? . . .)

(c) You should find several. How can you know when you have them all?

Investigation 1

Do not expect to find a **complete** answer. If you explore the problem sensibly, you should begin to notice lots of curious patterns. One way of starting is to make a list like this:

N	All numbers which are equal to N times the sum of their digits
1	1, 2, 3, 4, 5, 6, 7, 8, 9
2	.
3	.
4	.
.	.

It is a good idea to fill in this list at least as far as $N = 19$, 20, or 21 before trying to see exactly what is going on. (The strategies you use to compile this list efficiently and accurately, making absolutely sure that your list is complete for each value of N, are an important part of the exercise.) You may wish to check whether your table is complete by testing each number, up to say 200, to see whether it is divisible by the sum of its digits.

Investigation 2

There is no quick solution here, but one can do a lot better than simply testing all the possible expressions $a^b \times c^d$ in turn. (How many such expressions are there? More than six thousand!) For example, could we possibly have $a = 1$? If we could, then $c^d = a^b \times c^d = 1bcd$. How many powers c^d are there which lie between 1000 and 2000? What are they? Does either of them end with the two digits cd?

2

Near misses

1. In the multiplication 6 × 2 = 3 all the digits are correct, but they are in the wrong places! The sum should be 2 × 3 = 6 (or 3 × 2 = 6). In each of the next three multiplications all the digits are correct, but some of them are in the wrong places. Can you put them right?

    (a) 28 × 1 = 44   (b) 43 × 2 = 14   (c) 76 × 8 = 41

Are you sure there is only one correct answer each time?

2. (a) In these additions the digits are all correct, but they are in the wrong places. Can you put them right?

```
   1 0              3 2
   1 8 +            1 7 +
  -----            -----
  2 8 2            4 9 0
```

Are you sure that there is only one answer to each one?

    (b) In this division the digits are all correct, but they are in the wrong places. Can you put them right?

```
        5 7
   53 ) 2 9 4
```

How many different answers are there?

3. (a) Every digit in this multiplication has been copied down wrongly. But each digit is only one out. Can you find the original sum?

```
   1 6
     4 ×
   ---
   6 4
```

Are you sure your answer is the only possible one?

    (b) The next five are just like the one in part (a) except that some of them may have more than one answer. (If you think your answer is the only one, make sure you can explain why – even if a complete explanation is too complicated to actually write down.)

Every digit is one out. Can you find the original sums?

$$
\begin{array}{r} 2\ 5\ 2 \\ 6\ \times \\ \hline 1\ 5\ 1\ 2 \end{array}
\qquad
\begin{array}{r} 3\ 4 \\ 6\ \times \\ \hline 2\ 0\ 4 \end{array}
\qquad
\begin{array}{r} 6\ 7\ 8 \\ 5\ \times \\ \hline 1\ 2\ 4\ 5 \end{array}
\qquad
\begin{array}{r} 1\ 6\ 6\ 3 \\ 3\ \times \\ \hline 2\ 1\ 3\ 7\ 9 \end{array}
\qquad
\begin{array}{r} 2\ 6\ 8\ 3 \\ 3\ \times \\ \hline 2\ 6\ 1\ 7\ 9 \end{array}
$$

4. Here are some multiplications which have been copied down wrongly. But this time all you know is that each digit is *at most* one out. Try to find the original sums. (Where you think your answer is the only one, make sure you can explain why.)

$$
\text{(a)}\quad
\begin{array}{r} 7\ 8\ 9 \\ 6\ \times \\ \hline 2\ 3\ 4\ 5 \end{array}
\qquad
\text{(b)}\quad
\begin{array}{r} 3\ 7\ 5\ 3 \\ 3\ \times \\ \hline 1\ 1\ 2\ 5\ 9 \end{array}
\qquad
\text{(c)}\quad
\begin{array}{r} 7\ 8\ 4\ 2 \\ 4\ \times \\ \hline 3\ 1\ 3\ 6\ 8 \end{array}
$$

Investigation 1
(a)   The equation $12 \times 3 = 45$ can be corrected by changing exactly one digit. Which one?
(b)   What is the *smallest* number of digits which have to be changed in the equation $123 \times 4 = 567$ to put it right? How about the equation $1234 \times 5 = 6789$?

Investigation 2 (Do Question 3(a) first.)
How many multiplications of the form 'two digit number times one digit number equals two digit number' can you find which look correct (like '$16 \times 4 = 64$'), and in which every digit can be changed by one to give another correct multiplication?

Commentary

2.   (a)   For the first sum $(10 + 18 = 282)$ you want a 'carry' into the hundreds column, so one of the digits in the tens column will have to be an __ . And what must the hundreds digit be in the answer? The second sum $(32 + 17 = 490)$ is a bit harder. But you should be able to put one of the digits in its correct place straight away. And to get a 'carry' into the hundreds column, the tens column must contain either a __ or a __ .
   (b)   You will have to think quite hard about this one. (In this question, and many that follow, you should be prepared to work systematically with a number of alternative possibilities. For example, you may discover that a particular digit has to be either a 2, a 3, or a 4, but you don't know which! Your working can then split into three parts where you consider first 2, then 3, then 4. Be careful to examine each possibility in a systematic way.) One good place to begin here is to think about the very first step of the division.

$$
\begin{array}{r} ? \\ \cdots\overline{\smash{)}\ \cdots} \\ \cdot\ \cdot \end{array}
$$

The smallest number you could possibly be dividing by is 'twenty-something', and the largest number you could be dividing into is 'ninety-something'. So there are at most three possibilities for '?'. One of these can easily be seen to be impossible. You can then consider each of the two remaining possibilities in turn. (If you are clever, you can reduce the work by half!)

3.   (a)   Each digit is only one out. So the '1' in the tens column should be a __ . The answer must be a two digit number, and the multiplier '4' is only one out. So what should the multiplier be?

(b)   In the first one you can start at either end.

(i)   If you begin by looking at the left hand digits, you will notice that the thousands digit is only one out. So what should it be? There is only one way to change the hundreds digit '2' and the multiplier '6' to get this many thousands in the answer.

(ii)   But if you like, you could also start by looking at the units column. The '2' in the answer should be a '1' or a '3'. Only one of these can be obtained by changing the multiplier '6' and the '2' above it. This sorts out the units column. Similar ideas should allow you to complete the first one, and to do all the others.

4.   (a)   The units column is clearly wrong. However it is not clear whether it should be

```
   . . 9          or      . . 9          or      . . 8
     5 ×                    6 ×                    7 ×
  . . . 5                . . . 4                . . . 6
  ───────                ───────                ───────
      4                      5                      5
```

So perhaps you should start by looking at the other end. What is the least possible 'carry' from the hundreds to the thousands column? This should tell you the correct digit in the thousands column of the answer. It should also rule out *one* of the three possible multipliers. You must then consider the two remaining possibilities for the multiplier.

(b), (c)   These two look perfectly correct as they stand. But you were told that they were copied down wrongly (though not necessarily *completely* wrongly).

Investigation 2

(i)   There are a lot more than you might expect. It may be a good idea to begin by finding all such multiplications of the form '** × 1 = **', then those of the form '** × 2 = **', and so on.

(ii)   Suppose we start with '$ab × 1 = ab$'. The multiplier '1' must increase to a '2'. But then the tens digit 'a' on the left hand side must change to an __ , whereas the tens digit 'a' on the right hand side changes to an __ . If the 'b' changes the same way on both sides, it must change to a '0' (Why?), so the original sum was '$*1 × 1 = *1$'. This gives one solution. In all other cases the 'b' changes differently on each side (Why?), so we get either:

$$(a - 1)(b + 1)' × 2 = '(a + 1)(b - 1)', \quad or$$
$$(a - 1)(b - 1)' × 2 = '(a + 1)(b + 1)'$$

Each gives one solution.

If we now consider '$ab × 2 = cd$', we may assume that the multiplier '2' *increases* to a '3' (Why?), and then proceed in a similar way: the tens digit 'a' on the left hand side must change to a __ , while the tens digit 'c' on the right hand side must change to a __ .   And so on.

3

Colouring maps

A mathematical map is any arrangement of lines which divide the plane into separate regions (or countries). The first mathematical map shown here looks like a real map (whose regions are 'American states' rather than 'countries'). But most mathematical maps, like the second map shown here, are not really maps *of* anything. There is another important difference between real maps and mathematical maps: mathematical maps have to cover the whole plane. So, though the first two maps shown here look like 'islands', we have to imagine the 'outside', or 'sea', as an extra region, or country. In the third map the arrowheads on the ends of the straight lines tell us to imagine these lines going on forever – so there are several 'outside regions'.

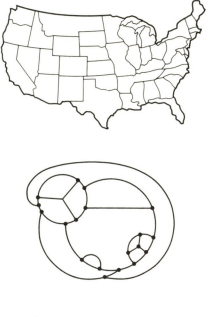

1. How many regions are there in each of the three maps shown above?

2. How many regions are there in each of these maps?

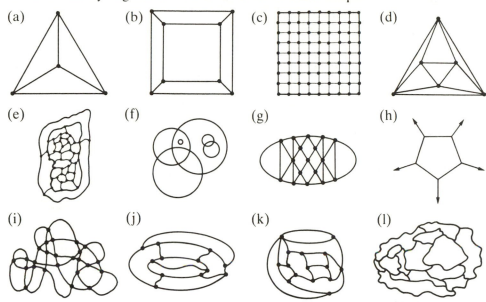

(a)   (b)   (c)   (d)

(e)   (f)   (g)   (h)

(i)   (j)   (k)   (l)

Suppose we want to colour the countries of a real map so that the different countries stand out. Then any two neighbouring countries must be coloured with different colours: any map which has been coloured according to this simple rule is said to be **properly coloured**. (Two countries which meet only at a point without sharing a stretch of common border are not strictly neighbours, and so can be coloured with the same colour.)

Instead of using coloured ink we often write numbers, 1, 2, 3, and so on, in the regions to denote different colours. The first map on the right has been properly coloured with two colours – black and white (the outside region has been coloured white). The second map cannot be properly coloured with two colours, but has been properly coloured with three colours.

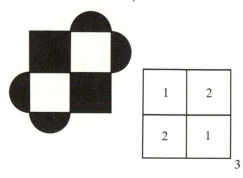

3. Which of these maps has been properly coloured?

(a)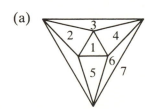

(b)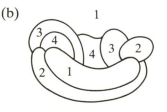

(c)

1	2	1
2	1	2
1	2	1

4. Which of the maps in Question 1 and Question 2 can be properly coloured with two colours? Which ones need three colours? Which ones need four colours? Which ones need five or more colours?

In 1852 a student at University College London suggested that *every* map could apparently be properly coloured using at most four colours, no matter how complicated the map might be. You may find this rather hard to believe! After all, the map in Question 2(a) shows that one sometimes needs four colours to colour even very simple maps properly. And you may have had a little difficulty finding a proper colouring of the map in Question 2(l) using just four colours, though the map could scarcely be called complicated. So one would expect some very large and complicated maps to need more than four colours. It is important to realise that what matters here is not whether you or I manage to colour a particular map properly using at most four colours at the second or third attempt, but whether it is always possible, for every map, to get away with using at most four colours. The fact that one can always get away with at most four colours was not proved until 1976.

Investigation 1

Two players take turns to colour the regions of a map – one region per go. The loser is the first player who has to use a fifth colour to avoid spoiling the proper colouring.

(a)   With the map shown here one player can win every time, no matter what the other player does. (Remember, the outside region also has to be coloured!) Which player can win every time? The one who goes first, or the one who goes second?

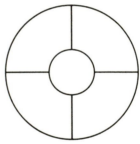

(b)   Can you construct another map which will allow the player who goes first to win every time?

Investigation 2

Every map can be properly coloured with either two, or three, or four colours. Can you discover an easy way of recognizing those maps which are so simple that they can be properly coloured with just two colours?

Commentary

1. Don't forget the 'outside' regions. The exact answer to the first one is obscured by the small scale of the map (and the fact that the Great Lakes count as regions). The answers to the other two are 12 and 11.

2. (h) The five lines with arrowheads are supposed to go on forever. How many 'outside' regions does this create?

3. Examine each one carefully. ((a) One region has been forgotten. (b) Two neighbouring regions have been coloured the same. (c) The outside region has been forgotten.)

4. They can all be properly coloured using at most four colours. But only just over a third of them actually *need* four colours: the rest can be properly coloured with either two or three colours. (It is very easy to think you have done some of these, such as (e), or (l), when you haven't. Check your colourings very carefully. If possible get a friend to check each one to see whether you have accidentally coloured neighbouring countries the same.)

Investigation 1
(b) If you want the simplest possible answer, you could just modify the map in part (a) ever so slightly by introducing one extra region without changing things too much. After that you might agree that the real challenge is to produce a map with at least seven regions, in which the first player can be sure of victory every time.

Investigation 2
(i) Of all the maps in Question 1 and Question 2, exactly five can be properly coloured with two colours. Which are they? Compare the five maps which can be properly coloured using just two colours with the ten which cannot. Do you notice anything which might help you to spot those maps which can, and those which cannot, be properly coloured with two colours? **Try hard to answer this question before reading any more of the Commentary!**

(ii) You may have noticed a good way of picking out some maps which could not possibly be properly coloured with two colours.

Suppose, for example, that a map contains *seven* countries A, B, C, D, E, F, G which all meet at a point. Could the map be properly coloured with just two colours? What is special about the number seven here?

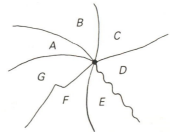

(iii) Part (ii) should have led you to a test which tells you that some maps cannot possibly be properly coloured with two colours. Go back and use your test on the fifteen maps in Question 1 and Question 2. Which ones does it rule out?

(iv) So a map which *fails* your test definitely *cannot* be properly coloured with two colours. However, it does not follow from this that every map which passes your test *can* be properly coloured with two colours. (It may still turn out to be true that every map which passes your test can in fact be properly coloured with two colours; but it is not true as a consequence of (ii) and (iii) above.) If you find this hard to understand, Extension 1 below should help. If you think you do understand, you may prefer to go on to (v) and come back to Extension 1 later.

Extension 1 (a) Which of the maps in Question 2 have exactly three countries meeting at each dot?
If a map has exactly three countries meeting at each dot, then it is easy to colour the countries round each dot properly using exactly three colours (since there are only three countries to colour).
(b) Suppose you wanted to know exactly which maps can be properly coloured with at most three colours. There is a simple-minded way of ruling out some maps which definitely *cannot* be so coloured: if the countries round some dot cannot be properly coloured with three colours, then neither can the whole map. The maps you found in part (a) all *pass* this simple-minded test: the three countries round each dot can certainly be properly coloured with three colours. Does it follow that the whole map can be properly coloured with three colours? Look back at each of the maps you identified in part (a) and see!

(v) So though any map which *fails* the test you found in part (ii) definitely *cannot* be properly coloured with two colours, Extension 1 above should make you think twice before assuming that every map which *passes* your test definitely *can* be properly coloured with two colours. Go back and check all the maps in Section 3 to see whether those maps which pass your test can in fact be properly coloured with two colours.
(vi) You are now faced with a problem. You have a test which seems to do more than you have a right to expect. Does it really work as it seems to? Could it in fact be true that any map which *passes* you test can *always* be properly coloured with two colours? And if it works, why does it work? Try it out on some maps of your own to help you decide whether you think it always works.
(vii) Suppose I have a map which passes your test, and I try to colour the countries with two colours – black and white – like this:
  (1) I start with any country A and colour it white.
  (2) I then have to colour all the neighbours of A black.
  (3) Next, every country which shares a stretch of border with one of these black countries has to be coloured white.
And so on. What could possibly go wrong? Can this really happen? If not, why not?

Extension 2 To prove that every mathematical map can be properly coloured using at most four colours, it turns out that one only has to worry about maps in which exactly three countries meet at every dot.
(a) Which of the maps in Question 1 and Question 2 have exactly three countries meeting at every dot?
(b) You now know a rule which tells you exactly when a map can be properly coloured with just two colours. Can any maps with exactly three countries meeting at every dot be properly coloured with just two colours? Can you find a rule which tells you exactly when such a map can be properly coloured using *three* colours?

4

Squares and cubes

1. The floor of a square hall is tiled with square tiles. Along the two diagonals there are 125 tiles altogether. How many tiles are there on the floor?

2. You are given a piece of wood measuring 8 cm by 18 cm. The wood is cut into two pieces roughly as shown here, and the pieces are then rearranged to make a square. Find the exact measurements.

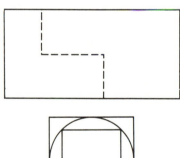

3. Here are two squares. One square is inscribed inside, the other is circumscribed about, a given circle. If the smaller square has area 3 square units, what is the area of the larger square?

4. The list of numbers

49,   4489,   444889,   44448889,   4444488889,   . . .

goes on for ever. What is the next number in the list? The first number 49 is a perfect square. Is the second number 4489 a perfect square? Is the tenth number in the list a perfect square? Which other numbers in the list are perfect squares?

5. Here are thirty two dots arranged in a 'hollow square'. In how many different ways can 960 dots be formed into a hollow square.

6. A hollow cubical box without a lid contains sixty four small wooden cubes which fill it exactly. How many of these small cubes touch the sides or the bottom of the box?

7. (a) Could 33∗∗6 be a perfect square? (The two missing digits do not have to be the same.)

(b) Could 301∗∗ be a perfect square?

8. Two identical squares overlap as shown, with one corner of one square at the centre of the other square. Find the area of the overlapping region.

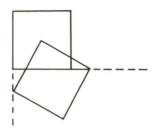

9. Some unit cubes are put together to make a larger cube. Then, some of the faces of this larger cube are painted. When the paint has dried, the larger cube is taken apart.

(a) Suppose it is then found that forty five of the unit cubes have no paint on any of their faces. How many faces of the larger cube were painted?

(b) What if exactly one hundred of the unit cubes have no paint on any of their faces? Can you say how many faces of the larger cube were painted?

10. One hundred and twenty five unit cubes are to be glued together to make a solid five by five by five cube.

(a) How many pairs of one by one faces are there to glue together?

(b) If it takes exactly one minute every time we glue two pieces together (using instant quick-drying glue), what is the most efficient way of assembling the big cube, and how long does it take?

11. (a) There are exactly eight numbers less than 100 whose squares have all their digits even. Find them all.

(b) Observe that $1^2 = 1$, so the square of 1 has all its digits *odd*. Which other squares have all their digits odd?

12. (a) How many different two dimensional shapes can be made by fitting four unit squares together edge to edge?

(b) A carpenter makes sets consisting of one shape of each kind, but using unit cubes instead of squares. He decides to sell each set in a five by two by two box or in a five by four by one tray. Which of these containers would you advise him to use?

Investigation 1

36 is the square of 6 and also ends in 6. 5776 is the square of 76 and also ends in 76. Which other numbers have the same property?

Investigation 2
When you multiply any number ending in 5 by another number ending in 5 the answer also ends in 5. Which other endings have a similar property?

Investigation 3
It is easy to find square numbers which end in a long string of zeros: $10^2 = 100$, $100^2 = 10000$, and so on.
(a)   Can you find square numbers which end in one 9? Can you find square numbers which end in two 9's? Can you find square numbers which end in a long string of 9's? What is the longest possible string of 9's you can have on the end of a square number?
(b)   Can you find square numbers which end in one 8? What is the longest possible string of 8's you can have on the end of a square number?
(c)   What about square numbers ending with a string of 7's? Or 6's? Or 5's? Or 4's? Or 3's? Or 2's? Or 1's?
(d)   If x is a digit between 1 and 9, which numbers of the form $xxx...x$ are perfect squares?

Commentary

1.   Can you see a connection between the number of tiles along one diagonal and the number of tiles along the edge of the square?

3.   No calculation is actually needed. (Try rotating one of the squares through 45°.)

4.   (i)   49 is obviously equal to 7^2. What is 4489 the square of? (*Never* use a calculator to answer a question like this without first trying to work out what the answer should be in your head!)
   (ii)   Now guess what 444889 is the square of. Then check to see whether your guess was correct.
   (iii)   What do you expect 44448889 to be the square of? Check to see whether your guess was correct. How can you be sure that this pattern will go on forever? Try to explain.

5.   There are at least ten different ways. Try to find them all. (Find a simple algebraic expression that must equal 960.)

6.   See Extension.

Extension   Suppose the box contains n^3 small wooden cubes. How many will touch the sides or the base of the box?

7. (a) What is the smallest possible number of the form 33**6? What is the largest such number? Are there any squares between these two numbers? (You should be able to work out 18^2 in your head:

$$18^2 = (20 - 2)^2 = 20^2 - (4 \times 20) + 4 = 324. \text{ So } 180^2 = 32400.$$

The square just above 33006 is therefore _____ . And the next square is _____ .)

(b) You should know that $17^2 =$ _____ , so $170^2 =$ _____ . The square just below 30100 is therefore _____ . And the next square is _____ .

8. No calculation is necessary.

> **Extension** With two identical squares the overlap is always the same, no matter how the top square is positioned, provided only that it has one corner at the centre of the bottom square.
> (a) Is the same thing true for two equilateral triangles?
> (b) For which other regular polygons is the same thing true?

9. (a) One good way of tackling this question is to shut your eyes, concentrate on a mental picture of a large cube, and try to do the whole thing in your head.

(b) Careful! There is more than one answer to this one.

10. (a) Each unit cube has _____ faces, so there must be _____ one by one faces altogether. How many of these land up on the outside of the five by five by five cube?

11. (b) It shouldn't take long before you think you know the answer here. But you may have to work out quite a lot of squares, preferably by hand, before you can see *why*. (If N is a number whose square N^2 has all its digits *odd*, then the units digit of N must be _____ . So you can concentrate on numbers which end in 1, 3, 5, 7, or 9.)

12. (b) One shape is different from all the others. Can you see which one? (If not, try shading the squares as if each shape were drawn on a chessboard.)

Investigation 1

Other than 0 and 1 there are just two single digit numbers with this property. What are they? How many two digit numbers are there with this property? What are they? And how many three digit numbers? Four digit numbers? Five digit numbers? . . . (If you use a calculator, or a computer, for some part of this investigation, think first what you expect to find. Afterwards try to explain what you *do* find.)

Investigation 2

One sensible beginning is to find all single digit endings first. (How many are there?) But there is no reason to stick to the units digit only. How many two digit endings are there with a similar property? How many three digit endings? . . .

Investigation 3

(d) This may be a little surprising, so it is worth persevering.

5

Reach for the stars

1. Some of the digits in these multiplications are missing. Can you find all the missing digits?

(a)
```
    * 6 *
      7 ×
  -------
  * 1 * 3
```

(b)
```
    * 7
    * * ×
  -----
  4 * 3
```

(c)
```
    6 *
    * * ×
  -----
  3 * 4
```

2. This time *all* the digits are missing. But you should still be able to work out what goes where.

   (a)   Can you replace the *'s with the digits 1, 2, 3, 4 using each digit once? How many different solutions are there?

```
        *
   * ) * *
```

In some of the other puzzles in this question it may not be possible to insert the digits as instructed; in others there may be more than one answer. So don't just guess! When there are solutions, make sure you find them all. And when you think there is no solution, try to explain why there is no solution.

   (b)   Can you replace the *'s by the digits 1, 2, 3, 4, 5 using each digit once? How many different solutions are there?

```
   * *
   * ×
  ----
   * *
```

   (c)   Can you replace the *'s by the digits 1, 2, 3, 4, 5, 6 using each digit once? How many different solutions are there?

```
         * *
   * ) * * *
```

   (d)   Can you replace the *'s by the digits 1, 2, 3, 4, 5, 6, 7 using each digit once?

```
   * * *
     * ×
  -----
   * * *
```

Long Division

When you divide 3703 by a small number like 7 (without a calculator) it is easy to work out the remainders at each stage in your head. The working is usually set out something like this:

$$7 \overline{)3\ 7\ {}^20\ {}^63}$$
$$5\ 2\ 9$$

But when we divide 3703 by a larger number, such as 23, (again without a calculator), it is traditional to set out the working slightly differently to help to make sure that we get the remainders right.

Instead of $23 \overline{)3\ 7\ {}^{14}0\ 3}$,   we write $23 \overline{)\begin{array}{c} 1\ .\ . \\ 3\ 7\ 0\ 3 \\ 2\ 3 \\ \hline 1\ 4\ 0 \end{array}}$

We then divide 23 into 140 in the same way. So the completed division would look like this:

$$23 \overline{)\begin{array}{c} 1\ 6\ 1 \\ 3\ 7\ 0\ 3 \\ 2\ 3 \\ \hline 1\ 4\ 0 \\ 1\ 3\ 8 \\ \hline \ 2\ 3 \\ \ 2\ 3 \end{array}}$$

3. Some of the digits in these long divisions are missing. Can you find all the missing digits?

(a)
```
         * * *
   ** ) * * * * 5
        * 0
       -----
        * * *
        4 * *
       -----
          * *
          * *
         -----
```

(b)
```
         * 5
   ** ) * * * *
        * *
       -----
        * *
        * 5
       -----
```

(c)
```
         * * *
   ** ) * * * *
        * 5
       -----
        * * *
        * * 9
       -----
```

(d)
```
         3 *
   ** ) * * * 3
        * *
       -----
        * *
        * *
       -----
```

4. (a) One of the three digits in this division is wrong. Which one is it? And what should it be? Can you find all the missing digits?

```
          7 *
      _____
  *5 ) * * *
        * *
       _____
        * * *
        * * 8
```

(b) Here are three more divisions like the one in part (a). In each division one of the given digits is wrong. Work out which one is wrong and what it should be. Then find all the missing digits.

```
         6 *                    7 * *                       * 2 *
      _____              _____                  _____
  *5 ) * * *           *6 ) * * * *              *7 ) * * * * *
        * *                  * *                          * *
       _____                ____                        ____
        * * *                * * *                        * *
        * * 8                * * 7                         * 7
                                                          _____
                                                          * * *
                                                          * * 2
```

5. You are told just one of the digits in each of these long divisions. Can you work out all the others?

```
              * * *                              * * 8 * *
          _____                         _____
(a)  *7 ) * * * * *        (b)  *** ) * * * * * * * *
            * *                              * * *
           ____                             _____
            * *                              * * * *
            * *                              * * *
           ____                             _____
              * *                             * * * *
              * *                             * * * *
```

6. This time all the digits are missing. But you should at least be able to work out what number we are dividing by.

```
                          * * * * * *
                       _____
      ** ) * * * * * * * * *
            * *
           ____
            * * *
            * *
           ____
            * * *
            * *
           ____
              * *
              * *
             ____
              * * *
              * * *
```

*7. (a) Can you replace the *'s with the digits 1–9 using each digit once only? How many different solutions are there?

```
      * * * *
            * ×
      _____
      * * * *
```

(b)   Can you replace the *'s
with the digits 1–9 using each digit
once only? How many different
solutions are the there?

Commentary

2. (a)   You may find it easier to use the fact that:

$$a \overline{)\, b\ c\,}^{\,d} \quad \text{is another way of writing} \quad \begin{array}{r} a \\ d \times \\ \hline b\ c \end{array}$$

You can then use logical reasoning, not guesswork, to work out which digits go
where. (If 'a' and 'd' have to be 1, 2, 3, or 4, what is the largest possible carry to the
tens column? So what must 'b' be? Could 'a' and 'd' both be even?)

Extension   The numbers in Question 2(a) are obviously meant to be written in
base 10. However, the problem has a solution in exactly one other base. Can
you find it?

(b)   Think about the units column. What goes wrong
if for the multiplier 'c' we have $c = 1$ or $c = 5$? What
goes wrong if $b = 1$ or $b = 5$? Could $e = 5$? Could
$e = 1$?

$$\begin{array}{r} a\ b \\ c \times \\ \hline d\ e \end{array}$$

Extension   This problem has a solution in exactly two other bases. Can you
find them?

(c)   As in part (a) we can rewrite the division as a
multiplication. The commentary on part (b) should then
tell you a lot about the digits in the units column.
Could $a = 1$? Could $d = 5$?

$$\begin{array}{r} a\ b \\ c \times \\ \hline d\ e\ f \end{array}$$

(d)   This is a bit tougher than the first three, but
exactly the same kind of reasoning should still work. You
should be able to show (as in part (b)) that c, d, g cannot
be 5, and that c, d cannot be 1. Observe that there is no

$$\begin{array}{r} a\ b\ c \\ d \times \\ \hline e\ f\ g \end{array}$$

carry from the hundreds to the thousands column, and that e is at most 7. So you can
deduce that $a = 3$ and $d = \underline{\ \ }$, or $a = 2$ and $d = \underline{\ \ }$, or $a = 1$. Now try each of
these three possibilities in turn.

3. (a)   Which two *'s have to be 5? What does this tell you about the units digit of
the number you are dividing by? (How large must it be?) And what does the 0 tell
you?

(b)   You are told that '5 times the two digit divisor "ab" equals a two digit
number ending in 5'. This should tell you what the tens digit 'a' has to be. It should
also tell you quite a lot about the units digit 'b'. At the first stage of the division you
get 'a three digit number minus a two digit number leaves a one digit number'.

This tells you a lot more than you might have thought. (Think about it!) In particular, it rules out all except two of the possible divisors '*ab*'.

   (c)   What does the 5 tell you? (The 9 shows that the units digit '*b*' of the number '*ab*' you are dividing by cannot be a 5.) This should tell you the tens digit of the number '*ab*' you are dividing by.

   (d)   This is like part (b).

4.   (a)   (i)   There is no way of saying straight away which digit is wrong. All you can say is that certain pairs of digits cannot both be right.

   (ii)   Once you discover which digit is wrong, rub it out. One of the two correct digits should tell you the tens digit of the number you are dividing by. The rest requires a little more thought, but is not too hard.

   (b)   If you managed the one in part (a), then you should not have too much trouble with the first two here. (They are slightly harder, but not really different.)

5.   (a)   Look at the first step: '*bcd* − *gh* = *i*'. What does this tell you about *b*, *c*, and *g*? Then '*s* × *a*7 = *gh*' leaves just *two* possibilities for *a*. You have to decide which of these two possibilities fits the other facts.

```
              s t u
      _____
  a7 )b c d e f
       g h
       _____
         i j
         k m
         _____
           n p
           q r
```

   (b)   The first step '*defg* − *lmn* = *pq*' should tell you the value of *d*, *e*, and *l*. The second step '*pqrs* − *tuv* = *wx*' should tell you the value of *p*, *q*, and *t*. (You should then be able to work out *f* and *m*.) Then '8 × *abc* = *tuv*' should tell you what *a* has to be. And if you think carefully about '*lmn*' and '*tuv*' you should be able to find *b*. The rest is left to you.

```
                * * 8 * *
       _____
  abc )d e f g h i j k
        l m n
        _____
        p q r s
          t u v
          _____
          w x y z
          . . . .
```

6.   The ideas which were so effective in Question 5(a) should help you to show fairly quickly that there are *at most three* possibilities for the number we are dividing by. You must then look for other clues to rule out two of these.

Alternative   Here is a slightly easier version of essentially the same problem.
What number are we dividing by this time?

```
              * * * * *
       _____
  ** )* * * * * * * *
       * *
       _____
       * * *
         * *
         _____
          * *
          * *
          _____
          * * *
          * * *
          _____
```

7.   (a), (b)   These are both quite hard. The first has exactly two solutions; the second has just one solution.

6

Prime time: one

These problems form a set and should be tackled in the order in which they appear. Some questions may look similar to others, but that is no reason to skip them: they may not be quite the same! Your task is not just to find numbers of the kind required for each problem, but to think about what you find, to draw conclusions, and finally to explain as clearly and as simply as you can what you discover. (By the way, '1' is not a prime number.)

1. 2 and 3 differ by one and are both prime numbers. What is the next such pair?

2. 3 and 5 differ by two and are both prime numbers. What is the next such pair? How many pairs are there like this?

3. 3, 5, and 7 go up in steps of two and are all prime numbers. What is the next such triple?

4. Find a prime number which is one less than a perfect square.

5. Find another prime number which is one less than a perfect square.

6. Find a prime number which is one more than a perfect square.

7. Find another prime number which is one more than a perfect square. How many prime numbers are there like this?

8. Find a prime number which is one less than a perfect cube.

9. Find another prime number which is one less than a perfect cube.

10. Find a prime number which is one more than a perfect cube.

11. Find another prime number which is one more than a perfect cube.

12. What can you say about n when $2^n - 1$ is a prime number?

13. What can you say about n when $2^n + 1$ is a prime number?

14. Find a prime number of the form $3^n - 1$.

15. Find another prime number of the form $3^n - 1$.

16. Find a prime number of the form $3^n + 1$.

17. Find a prime number of the form $4^n - 1$.

18. Find another prime number of the form $4^n - 1$.

19. What can you say about n when $4^n + 1$ is a prime number?

20. (a) What can you say about m and n when $m^n - 1$ is a prime number?

(b) What can you say about m and n when $m^n + 1$ is a prime number?

Investigation
(a) Find the first prime number which is exactly 9 less than a perfect square. How big is the next one?
(b) Find a prime number which is 8 more than a perfect cube.

Commentary

This set of problems is unlike other sections. The questions may look harmless enough, but the experience of tackling and solving them should bring out, in a very clear and concise way, the three phases of any mathematical investigation: the initial exploratory phase, the gradual emergence of a guess which one has good reason to believe is correct, and the final stage where one tries to prove that this guess is indeed correct. (These three phases are not usually as separate as they are made to appear here!)

I would argue that *most* mathematical investigations in the classroom should be of this structured kind (which is currently the *least* common). Teachers inclined to agree should be able to devise problem sequences of their own in the same spirit.

2. How many such pairs can you find?

3. In Question 1 you probably realised that, when two numbers differ by one, one of them is bound to be *even*: in other words, one of them is bound to be *a multiple of 2*. In Question 3 the distinction between even and odd is no use at all: if three prime numbers go up in steps of two, they must all be odd! But a related idea will help. Can you see what it is?

5. Search really thoroughly. Then stop and think. Finally try to explain!

7. How many prime numbers of this kind can you find?

9. Follow the advice in the Commentary on Question 5.

11. Follow the advice in the Commentary on Question 5.

> **Extension** What can you say about prime numbers which are one less than a fourth power? What about prime numbers which are one more than a fourth power?

12. (i)   Look at the first ten numbers of the form $2^n - 1$. What do you notice about the value of n when $2^n - 1$ is a prime number? Can you show that n itself has to be 'special'? (You should be getting the message by now that general arithmetical questions of this kind are best resolved using 'generalized arithmetic': that is, 'algebra'! Suppose n can be written as a product $n = a \times b$. Can you use a little algebra to factorize $2^{ab} - 1$?)

(ii)   Now show that $2^{11} - 1$ is not a prime number.

13. (i)   Replace $2^n - 1$ by $2^n + 1$ in part (i) of the Commentary on Question 12.

(ii)   Show that $2^{32} + 1$ is not a prime number.

15.   What sort of a number is 3^n? So what can you say about $3^n - 1$?

18.   What has this to do with Question 12?

19.   What has this to do with Question 13?

You should be able to give complete explanations of everything you find in this section except for your answers to the supplementary questions at the end of Problems 2 and 7. These are both very interesting questions.

Pairs of prime numbers which differ by two are called 'twin primes'. Prime numbers occur less frequently among larger numbers than among smaller numbers. But the prime numbers don't just 'spread out'. No matter how far you go it seems that every so often you will come across a pair of prime numbers that are as close together as they could possibly be. There are 1224 pairs of twin primes less than 100,000, and 8164 pairs of twin primes less than 1,000,000. But as yet, no one has been able to prove that there are in fact *infinitely many* pairs of twin primes.

Another difficult unsolved problem concerns prime numbers of the form $n^2 + 1$. There are some obvious restrictions on the value of n if $n^2 + 1$ is to be a prime number: for example, apart from $2 = 1^2 + 1$, n must be even. But though surprisingly many numbers of the form $n^2 + 1$, such as

2, 5, 17, 37, 101, 197, 257, 401, 577, 677, 1297, 1601, . . . ,

do turn out to be prime numbers, we still do not know for sure whether infinitely many prime numbers are of this form.

7

Cunning c*lc*l*t**ns

Do not confuse the c*lc*l*t*rs in this section with real calculators. Unlike real calculators, **c*lc*l*t*rs behave strictly according to the laws of arithmetic**. In particular, the operations +, −, ×, ÷ can only be used to combine two numbers. Meaningless sequences of instructions, such as

⊟ ②, or ÷ ÷, or ② ⊠ ② ⊠, or ② ⊠ ⊟ ⊟,

which are accepted (and acted upon) by many real calculators, cause permanent damage to the sensitive internal circuitry of any c*lc*l*t*r.

1. There is one easy way of making 15 using the numbers 1, 2, 3, 4, 5, once each in their natural order, namely:

$$1 + 2 + 3 + 4 + 5 = 15.$$

How many other ways can you find of getting the same answer (15) still using only the numbers 1, 2, 3, 4, 5, once each in their natural order?

2. Can you get every number from 0 to 29 using the numbers 1, 2, 3, 4, once each in their natural order? (You are allowed to combine 1, 2, 3, 4 using +, −, ×, ÷, √, and by forming powers. You are *not* allowed to put 1, 2, 3, 4 together to make numbers like 123 or 34.)

3. Which of the numbers 0–20 can you get in the display of this curious c*lc*l*t*r if you have to press the ④ button exactly four times for each number? What is the smallest number which *cannot* be obtained by pressing the ④ button exactly four times? (This time you *are* allowed to put 4s together to make 44, or 444, or even 4444.)

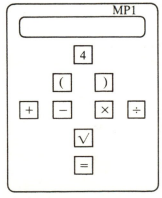

4. To get a 1 in the display of this c∗lc∗l∗t∗r you have to press the ② button at least twice: for example, ②, ÷, ②, ⊟. To get 2 you only need to press the ② button once. For some numbers N you have to press the ② button at least N times to get N in the display.
Which numbers N can you get in the display by pressing the ② button fewer than N times?

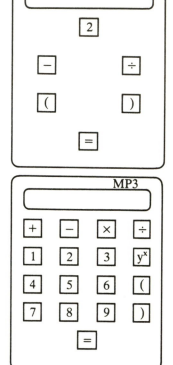

5. (a) What is the biggest number you can get on the c∗lc∗l∗t∗r shown here if the only *digit* buttons you are allowed to press are ① and ② *once each* (in any order)?

(b) What is the biggest number you can get if you are allowed to press the y^x button at most once, and the only digit buttons you are allowed to press are ①, ②, ③ once each (in any order)?

(c) What is the biggest number you can get if you are allowed to press the y^x button at most once, and the only digit buttons you can press are ①, ②, ③, ④ once each (in any order)?

(d) What is the biggest number you can get if you are allowed to press the y^x button at most once, and each of the digit buttons ① to ⑨ exactly once each?

∗(e) What happens in parts (b), (c), (d) if you are allowed to press the y^x button as often as you like? Would your answers still be the same as before?

Investigation 1 (Do Question 3 first.) On the c∗lc∗l∗t∗r shown here you can *choose* which of the digits 1–9 you put on the blank button. Which digit would it be best to choose if you still have to press this button exactly four times to get each number?

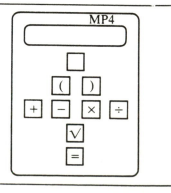

Investigation 2
Which of the numbers 0–100 can you get using the four digits of the current year (such as 1, 9, 8, 7) once each in their natural order, if you are allowed to form combinations involving $+$, $-$, $\times$, $\div$, $\sqrt{}$, and powers, or put digits together to make two-digit, three-digit, or four-digit numbers? Which other operations do you need to fill in the missing numbers – if any?

Commentary

1. There are at least three fairly easy ones, and several others requiring more ingenuity.

2. You should be able to do all except 17, 18, 19, 22, 26, 27, and 29 without using the $\boxed{\sqrt{}}$ button. All but one of these seven numbers can be got in the required way.

3. You should be able to do all except 4, 11, 13, 14, 18, and 19 without using the $\boxed{\sqrt{}}$ button. All but one of these six numbers can be got in the required way.

4. You can obviously get 22, or 20 ($= 22 - 2$), or 111 ($= 222 \div 2$), by pressing the $\boxed{2}$ button relatively few times. But these are very special numbers and you should not let them lead you astray. **You want an idea that will work for as many numbers as possible.** You clearly cannot get 3 in the display by pressing the $\boxed{2}$ button twice only. Can you get 4 in the display by pressing the $\boxed{2}$ button three or fewer times? Can you get 5 in the display by pressing the $\boxed{2}$ button four or fewer times? What about 6? Or 7? (Remember the sentence in bold type!)

5. (b) Which is biggest: 321, 2^{31}, or 3^{21}?

(c) Which is biggest: 421^3, 43^{21}, 21^{43}, 4^{321}, 3^{421}, or 2^{431}?

(d) Look carefully at your answers to parts (b) and (c).

(e) Which is bigger: $3^{(2^{4^1})}$ or $3^{(4^{2^1})}$? $4^{(3^{2^1})}$ or $2^{(4^{3^1})}$?

Investigation 1
It all depends what you mean by 'best'! Suppose you take this to mean: 'Which digit allows you to get all numbers from 0 up to N without any gaps and with N as large as possible?' Then you know from Question 5 what happens with a 4 on the blank button. If the blank button had a 1 on it, could you get a 5 in the display? What if the blank button had a 2 on it. Could you get a 7 in the display?

Investigation 2
One popular extra operation which I have omitted is the factorial sign '!'. For example, $9! = 9 \times 8 \times 7 \times 6 \times 5 \times 4 \times 3 \times 2 \times 1$, so $1 + (9! \div 8!) + 7 = 17$. It is best to try to get as many numbers as possible *without* introducing extra operations. (For 1, 9, 8, 7 you should be able to do three quarters of the numbers 0–100 without any extra operation.)

8

Word sums

1. (a) In the sum on the right
the words stand for numbers. Each
of the letters F, O, U, R, N, E, I,
V stands for one of the digits 0–9.

```
F O U R
  O N E +
F I V E
```

Different letters stand for *different* digits. And the first digit of a number
is never zero. (So F and O cannot be zero.) Can you replace the letters by
numbers to get a correct sum?

(b) Here are two more word-sums like the one in part (a). Can you
replace the letters by numbers to get a correct sum?

```
O N E          T W O
O N E +        T W O +
T W O          F O U R
```

The first one has sixteen different solutions. See how many you can find.
The second one has eight different solutions. How many can you find?

2. If you try making up word-sums like these for yourself, you may find
that they often look promising but don't actually work! When you come to
have a go at harder problems (like the ones in Question 4) it is important
to be able to choose a good place to begin. One way of learning to 'spot'
good starting points is to try to explain very simply why certain word-sums
can't be done. Most of the word-sums in this question can't be done. Find
the ones which can't be done and explain why they are impossible.

```
(a)  F I V E          (b)  T H R E E          (c)  F O R T Y
       T W O +               N I N E +               F O R T Y +
     S E V E N             T W E L V E             E I G H T Y
```

(d)
```
  E I G H T
    F O U R +
  T W E L V E
```

(e)
```
  E L E V E N
        O N E +
  T W E L V E
```

(f)
```
  F O U R
  F I V E +
  N I N E
```

(g)
```
  T W E L V E
    E I G H T +
  T W E N T Y
```

(h)
```
  T H R E E
    F I V E +
  E I G H T
```

(i)
```
    T H I S
    O N E 'S +
  A W F U L
```

3. Can you work out what T must be in this one?
```
  T H I S
      I S +
  H A R D
```

4. The most satisfying word-sums are those that can be shown to have just one solution by elementary reasoning. Most of the word-sums in this question are of this kind. They are harder, but more interesting, than the previous ones.

(a) How many different solutions are there to this one?
```
  H O C U S
  P O C U S +
  P R E S T O
```

(b) Find all possible solutions to each of these.

```
  S C A N
  T H E S E +
  D I G I T S
```

```
  C R O S S
  R O A D S +
  D A N G E R
```

```
  B E E R
  B E E R +
  D R U N K
```

```
    T E N
    T E N +
  F O R T Y
  S I X T Y
```

```
  S E V E N
  S E V E N +
    S I X
  T W E N T Y
```

```
  S E V E N
  T H R E E +
    T W O
  T W E L V E
```

5. How many different solutions are there to this one? (Carl Friedrich Gauss was perhaps the greatest mathematician who ever lived. He was born in the year 1777 in the town of Braunschweig, and died in 1855 in Göttingen.)
```
  1 7 7 7
  1 8 5 5 +
  C A R L
  G A U S S
```

6. To end up, here are three with a seasonal flavour. The first has just one solution, the second has two solutions, and the third has four solutions.

(a)
```
        A
  M E R R Y +
    X M A S
  T U R K E Y
```

(b)
```
        X M A S
  A) H A P P Y
```

(c)
```
        X M A S
  A) M E R R Y
```

Commentary

1. (a)   There are five hundred and twenty eight different solutions to this one so you shouldn't have too much trouble finding one! The aim of this first question is to make sure you understand the 'rules', namely:

   (i)   that each letter stands for just one of the digits 0–9, (so if a letter occurs more than once in a sum, it must stand for the same digit each time)

   (ii)   different letters stand for different digits

   (iii)   no number starts with a zero

In the sum in part (a), the two F's must stand for the same digit. There is no carry from the hundreds to the thousands column. This means that the letter O is at most 4. You should now be able to find lots of solutions.

(b)   Look at the units column. What sort of digit must O be? There are in fact just two possible values for O. Can you find them? (What would go wrong, for example, if O were equal to 6?) For each value of O the word-sum has eight different solutions.

$$\begin{array}{r} O\ N\ E \\ O\ N\ E\ + \\ \hline T\ W\ O \end{array}$$

2. (a)   What is the most that could be carried to the ten thousands column? So what is S? And what must F and E be? Now use this value of E in the units column. What goes wrong?

(b)   What must T in the hundred thousands column be? Now look at the ten thousands column.

(c)   What does the units column tell you about Y? And what does the tens column tell you about T?

(d)   How many different letters are there in this sum?

(f)   There are several solutions.

(g)   The left hand columns hold the key.

(h)   In the units column E + E equals T or T + 10. And in the ten thousands column we must have had a carry from the thousands column (since T and E have to be different). Hence E = T + 1. So what must T and E be? Now look at the tens column.

3.   The T and H in the thousands column have to be different. So there must be a carry from the hundreds column. This is only possible if the H in the hundreds column is large enough to combine with a carry from the tens column to produce a carry into the thousands column. So what must H be?

4. (a)   (i)   Look first at the ten thousands column. What is the largest possible carry to the hundred thousands column? So what is P?

   (ii)   P must have exactly this value in the ten thousands column as well. This leaves two possible values for H. What are they?

   (iii)   Now look at the units column. What sort of a number must O be? If H = 8, there has to be a carry from the thousands to the ten thousands column. So there is only one possible value for O. What is it? Use this to find S. Then show that the hundreds column goes wrong.

   (iv)   You are now left with just one possible value for H. Use this to find R and the two possible values for O. Then try each value of O in turn.

(b)   If you understood how to solve the word-sum in part (a), you should be able to tackle these. (One has two solutions, and the rest have just one solution.)

5.   How many solutions would you expect? Gauss may have been unique, but he was also worth at least twice as much as any other mathematician! (There is an excellent short biography *Carl Friedrich Gauss* written by Tord Hall, and published by MIT Press in 1970.)

Word-sums like the ones in this section are not in themselves mathematically significant. But they exhibit, on a very elementary level, many important features of mathematical reasoning. For example:

(1) at first sight it is not at all clear how hard a given word-sum will be to solve
(2) there is no easy way to tell in advance whether or not a given word-sum has a solution at all or if it has, whether there is only one
(3) there is no obvious way of getting started, and at each stage one has to use one's intelligence to decide what to do next
(4) a successful solution nearly always depends on following through a systematic approach

There are lots of variations on the word-sum idea: they may be in other bases, or other languages, or include some additional pertinent information. Here is a selection of examples; (some may have more than one solution).

```
  E I N S                 Z W E I                   U N I T E D
  E I N S +               Z W E I +                   S T A T E S +
  Z W E I   base 6        V I E R   base 6          A M E R I C A   base 11


  E I  N S                  H A V  E                   E I N S
  E I  N S +                S O M  E +                 E I N S +
  Z W E I                   S U M  S                   E I N S
  V I  E R   base 8             T  O                   E I N S
                           S O L  V E                  V I E R   base 8


      U N                      U N                   T R O I S
  D E U X +                D E U X +                 T R O I S +
  D E U X                  D E U X                       S E P T
  D E U X                  D E U X                       S E P T
  D E U X                  D E U X                   V I N G T
  N E U F                  D E U X
                          O N Z E                   (7 divides SEPT)


  T W E N T Y
  T W E N T Y +
  T H I R T Y
  S E V E N T Y
```

(30 divides THIRTY)

Finally a personal favourite: $\sqrt{\text{PASSION}}$ = KISS

9

On your marks

1. You have an unmarked piece of wood which is exactly 12 units long. By making eleven marks on it – one mark 1 unit from the left hand end, the next one 2 units from

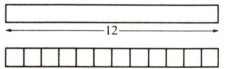

the left hand end, and so on – you could make a ruler which could measure each whole number of units, from 1 unit up to 12 units, *directly*. But this way of marking your piece of wood is very 'wasteful' in the sense that there would then be *twelve* different ways of measuring 1 unit *directly*, *eleven* different ways of measuring 2 units *directly*, and so on. What is the smallest number of marks you need to make on your piece of wood in order to be able to measure directly each whole number of units from 1 unit up to 12 units?

2. You have a circular piece of transparent plastic whose centre is marked. If you mark all the hour points (like a watch), then you would have a protractor which could measure any multiple of 30° directly. But this way of marking out your protractor is very wasteful in the sense that there would then be twelve different ways of measuring 30° directly, twelve ways of measuring 60° directly, and so on. What is the smallest number of radial marks you need to make on your blank piece of plastic in order to be able to measure each multiple of 30° *directly*?

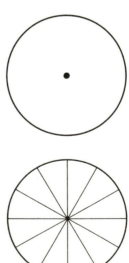

3. If you have a 1 cm piece of wood, you do not need to make any marks in order to be able to measure directly each whole number of centimetres up to 1 cm. If you have a 2 cm piece of wood, you obviously have to make at least one mark in order to be able to measure directly each whole number of centimetres up to 2 cm. Try to find the smallest number of marks needed on blank pieces of wood of lengths 3 cm, 4 cm, 5 cm, and so on up to 13 cm (say) to measure directly each whole number of centimetres.

Length of blank piece of wood	1 2 3 4 5 6 7 8 9 10 11 . . .
Smallest number of marks needed	0 1 . . .

4. (a) The first nine entries in the table in Question 3 lead you to expect that the tenth entry should be a three. But it isn't! Three marks plus the two ends certainly produce ten *pairs* of marks, but it is not possible to mark the ruler so that the ten distances they measure are all different. Can you explain why not?

(b) Part (a) suggests that there is no easy way of predicting exactly how many marks will be needed for an efficient ruler. But you *can* say something. What can you say about the number of marks needed for a 21 cm efficient ruler? And what about a 20 cm efficient ruler?

(c) Go back to the explanation you gave in part (a). Can you use the same idea to show that the only 'one hundred per cent efficient' rulers are those of length 1 cm (with no marks), 3 cm (with one mark), and 6 cm (with two marks)? All other 'efficient' rulers are bound to have some lengths which can be measured in more than one way.

As the length of the ruler increases it gets harder and harder to predict *exactly* how many marks will be needed to make the most efficient ruler. The next question explores just how efficient an efficient ruler can be.

5. When you choose where to put the marks on your 'efficient' rulers, you naturally try to avoid producing different pairs of marks which measure the same length (the two ends of the piece of wood being used as invisible marks).

If you make one mark 1 cm from the left hand end of a 3 cm piece of wood, then the three pairs of marks measure three different lengths, so no pairs are 'wasted'.

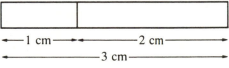

(a)   The most efficient way of marking a 4 cm piece of wood uses two marks. (There are two different ways of marking such a ruler.) If you use the ends as invisible marks, how many *pairs* of marks are there on such a ruler? Which lengths can be measured directly in more than one way? So how many pairs are wasted?

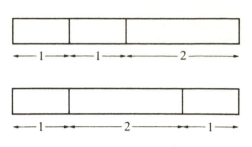

(b)   For a 4 cm ruler marked as in part (a), there are six pairs but only four lengths. So two of the pairs are wasted. Go back to the table you made in Question 3 and work out the number of different pairs on each of your 'most efficient' rulers, and the number of different lengths measured. Make a table like this one.

Length of ruler	1	2	3	4	5	6	7	8	9	10	11	12	13
Number of different lengths measured													
Number of marks required													
Number of pairs of marks (using ends)													
Number of 'wasted' pairs													

You should by now appreciate the difficulty of predicting exactly how many marks are needed each time. The final question in this section examines the same phenomenon from a rather different point of view: this time we fix the number of marks rather than the length of the ruler, and try to find the length of wood required if we are determined to avoid getting repeated measurements.

6.   In Question 1 you started with a 12 unit piece of wood and tried to find the smallest number of marks needed to make a ruler which could measure *directly* each whole number of units up to 12 units. In this question we work the other way round. We fix in advance the number of marks we are going to make, and then try to find the shortest piece of wood which can be marked with this fixed number of marks so that all the direct measurements between pairs of marks (including the ends) are *different*.

(a)   Suppose we wish to make *two* marks. If different pairs of marks must measure different lengths, then you saw in Question 5(a) that a 4 unit piece of wood is too short. Is it possible to make two marks on a 5 unit piece of wood so that different pairs of marks always measure different lengths?

(b)   Make a table of your own like this one.

Number of marks to be made	1	2	3	4	5	6	7	8	9
Shortest length of wood for which different pairs of marks always measure different lengths	3	6							

Investigation

Suppose we allow 'hinged' rulers like the one shown on the right. This is hinged at C so that ACD, ACB, and BCD can all be laid out straight. Can you choose the lengths of the three 'branches' AC, BC, CD so that every length from 1 cm to 6 cm can be measured?

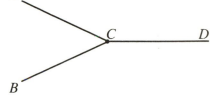

Try to choose the lengths of the 'branches' in the diagrams below to make efficient 'hinged' rulers. Then invent some efficient hinged rulers of your own.

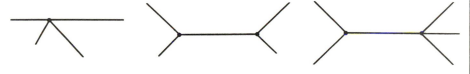

Commentary

1. (Do not read this note until you have found what you believe to be the smallest number of marks.) If you make no marks at all, then you can only measure one length directly – namely 12 units. Each time you measure a length you use a pair of marks. For this purpose, the two ends of the piece of wood count as 'invisible' marks.

(i)   Suppose we make just one mark on the piece of wood. How many different *pairs* of marks are there (using the ends as 'invisible' marks)? So what is the largest possible number of different lengths you could measure directly if you make just one mark?

(ii)   Suppose you make just two marks. How many different *pairs* of marks are there? So what is the largest possible number of different distances you could measure directly if you make just two marks?

(iii)   What is the least number of marks you could possibly get away with if you wish to measure *twelve* different lengths directly? You should now try to find a solution which uses precisely this smallest conceivable number of marks. (There are in fact five different ways of marking a 12 unit ruler with this smallest conceivable number of marks so that each length from 1 unit up to 12 units can be measured directly. Once you have found one way it is very worthwhile finding the other four.)

Extension Suppose you have a 12 by 12 sheet of transparent plastic. If you draw a grid of *eleven* equally spaced vertical lines and *eleven* equally spaced horizontal lines you would get an 'area measurer' which could measure directly any x by y rectangle, where x and y are any whole numbers up to 12. There would then be 144 ways of measuring a 1 by 1 square directly. What is the *smallest* number of lines you need to mark in order to measure every such rectangle directly?

2. The main difference between this question and the previous one is that a circle has no 'ends'. And as soon as you know you can measure $x°$ (say), you can automatically measure $(360 - x)°$ as well. **This means that you only have to worry about multiples of 30° up to 180°.** If you think about the problem for a while, you should discover that it is very like putting marks on a piece of wood — though not a 12 unit long piece of wood.

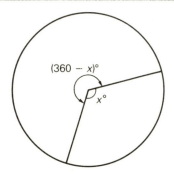

$(360 - x)°$

$x°$

3. For a 3 cm piece of wood you have to put at least one mark; and one mark will actually do. For a 4 cm piece of wood one mark cannot possibly do, since there would then be only three pairs of marks — using the ends — and these could only measure three lengths directly. So for 4 cm we must have at least two marks. And two marks will in fact do (there are two different solutions with two marks). Here are some of the entries in the table. (Note that the idea we used for 3 cm and 4 cm rulers tells us that a 10 cm ruler must have at least three marks. Is it possible to make a 10 cm ruler with just three marks?)

1	2	3	4	5	6	7	8	9	10	11	12	13
0	1	1	2		2		3			4		

4. (a) If it were possible to make an efficient 10 cm ruler with just three marks, every pair of marks would have to measure a different length. Since 9 cm has to be measured, there must be a mark 1 cm in from one end — say the left hand end. Since 8 cm has to be measured there must be a mark 2 cm in from one end (Why?), so this mark must be 2 cm in from the right hand end (Why?). We now have to make one further mark in the 7 cm gap without repeating any lengths. Can you see why this is impossible?

(b) You certainly know that at least five marks are needed (5 marks plus 2 ends give 21 pairs of marks). If it were possible to make an efficient ruler with just five marks, every pair of marks would have to measure a different length, so the first three marks would have to look like this (Why?). Can this ruler be completed to give an efficient 21 cm ruler with exactly five marks?

1	3		2

(c) Suppose your efficient ruler has length N cm and that every length from 1 cm up to N cm can be measured in exactly one way. Since $(N - 1)$ cm has to be measured, there must be a mark 1 cm in from one end – say the left hand end. Since $(N - 2)$ cm has to be measured, and we must not duplicate the 1 cm gap, there must be a mark 2 cm from the right hand end. Then $(N - 3)$ cm can be measured. (How?) But what about $(N - 4)$ cm and $(N - 5)$ cm? One of these forces us to make a mark 4 cm from the left hand end (Why?). The other then causes trouble whenever N is greater than 6.

6. (a) There are *six* pairs of marks on such a ruler. On the first ruler (with 'gaps' of length 1, 1, 2) both 1 cm and 2 cm can be measured in two different ways. On the second ruler (with 'gaps' of length 1, 2, 1) 1 cm and 3 cm can be measured in two different ways.

(b) Here are some of the entries in the table.

No. of marks made	1	2	3	4	5	6	7	8	9	10
Shortest ruler	3	6		17						

(i) If you make *three* marks (plus the ends), how many different pairs of marks are there? Can you show that it is impossible to put three marks on a 10 cm piece of wood so that different pairs of marks always measure different lengths? Can you show how to put three marks on an 11 cm piece of wood so that different pairs of marks always measure different lengths?

(ii) Try to fill in a few more slots in the table. You may then be interested to read Chapter 15 of Martin Gardner's book *Wheels, Life, and Other Mathematical Amusements*, published by W.H. Freeman in 1983.

Investigation
The lengths of the 'branches' of this hinged ruler can be chosen so that each length from 1 cm up to 15 cm can be measured *exactly once*. You should not expect other hinged rulers to be quite as efficient as this.

10

To and fro

1. (a) You have a 10 pint jug, a 7 pint jug, a tap, and no other container. Can you get exactly 9 pints?

(b) What if you have an 8 pint jug and a 5 pint jug? Can you get exactly 1 pint? Can you get exactly 6 pints? Can you get exactly 4 pints? Can you get exactly 7 pints?

(c) Can you get exactly 1 pint if you have a 7 pint jug and a 12 pint jug?

(d) Can you get exactly 1 pint if you have an 8 pint jug and a 13 pint jug?

2. (a) You have a 1 pint measure of milk and two cups – one $\frac{3}{8}$ pint cup and one $\frac{5}{8}$ pint cup. Can you divide the milk into two equal parts?

(b) You have a 12 pint jug of milk and two cans – one a 9 pint can and one a 4 pint can. Can you divide the milk into two equal parts?

3. The two trucks A and B are in a siding as shown. Can the engine E shunt A into B's position and B into A's position and finish up exactly where it started? (Each truck will just fit on the spur, so it is possible to shunt a truck onto the spur from one side and pull it off the other way. The engine is too big to fit onto the spur.)

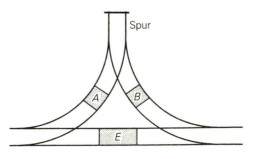

4. (a) You have to take a wolf, a goat, and some cabbages across a river. The only means of transport is a small rowing boat, just big enough to take you (the rower) and *one* other item of cargo on each trip. If you leave the wolf and the goat together (in your absence), the wolf will eat the goat. If you leave the goat together with the cabbages, the goat will eat the cabbages. How can you get everything safely across the river?

(b)   Four missionaries are on one side of the river and four cannibals are on the other side. Each group wishes to get to the other side and the only means of transport is a rowing boat big enough for three people. Only one missionary and one cannibal know how to row. The cannibals are likely to overpower the missionaries as soon as they outnumber them, whether in the boat or on either bank. Can the crossing be completed successfully?

(c)   What if there are five missionaries and five cannibals and the boat still carries only three people? What if there are six missionaries and six cannibals?

5.   A traveller runs out of money. But she still has a gold chain. She is offered accommodation at a hotel on the condition that she pays one link per day. Each morning she has to pay her one link. The woman stays until all the links have been used up.

(a)   Suppose the chain has seven links. She could simply cut the chain into seven separate links. What is the smallest number of links she would have to cut to do this?

(b)   If her chain has seven links, the lady could avoid cutting the chain into seven separate links provided she is allowed to take 'change' in the form of links which she has already paid. If this is done, what is the smallest number of links which she has to cut?

(c)   What is the smallest number of links that would have to be cut if the lady was allowed to take change and had a chain with *sixty* links?

Investigation   (Do Question 2(a) first.)

If you have an 8 pint jug of milk and two cans – one 3 pint can and one 5 pint can – then Question 2(a) shows you how to divide the milk into two equal parts. When is it possible to divide a given quantity of milk into two equal parts using two different-sized cans whose total capacity is equal to the given quantity of milk?

Commentary

1.   (a)   If you want 9 pints, which of the two jugs will it have to be in? The only obvious way of getting 9 pints in that jug is to start with 10 pints and pour exactly 1 pint into the other jug. The only way of pouring exactly 1 pint into the 7 pint jug is if it already contains exactly 6 pints. Can you see how to get exactly 6 pints in the 7 pint jug?

(b)   Yes. But how?

(c), (d)   If you look carefully at the way you got 9 pints in part (a), you may see a connection with the fact that 'three 10's minus three 7's equals 9'. This should suggest an easy way of getting 1 pint with a 12 pint jug and a 7 pint jug, or with an 8 pint jug and a 13 pint jug.

2. (a), (b) There is a subtle difference between Question 1 and Question 2. In Question 1 you can get more water from the tap, or pour it away (on the ground), as often as you like. In Question 2 you can't. Both parts can be done but I leave you to find your own solution. (The first part can be done in six pourings, the second in eight pourings.)

3. First shunt A way off to the right, and B way off to the left. Then think how to use the spur.

4. (a) Which two cargoes can be left together safely on the same bank in your absence? The other cargo is obviously the one you will have to keep you eye on all the time – even if it means taking it with you 'in the wrong direction' from time to time.

(b) (i) Where must the boat be to start with? On the first trip the boat may carry either one, or two, or three people. Decide which of these you are going to try. Then carry on from there, keeping a look out all the time to see whether you are about to run into trouble. If you get stuck, keep trying. Don't give up.

(ii) When you have found one solution, look at it very carefully and try to see how to get a second solution from it without doing any more work!

Extension (a) Suppose you have a boat which carries just *two* people at a time. Two missionaries (one of whom can row) and two cannibals (one of whom can row) could obviously use it to swap banks safely. Could three missionaries and three cannibals use it to swap banks safely? What about four missionaries and four cannibals?
(b) If you have worked through Question 4 then you should know what can, and what cannot be done with a boat which carries *three* people.
(c) How many missionaries and cannibals could swap banks safely if they had a boat which carries exactly *four* people?

5. (I have assumed that the chain has two ends. If you assumed that the chain was joined up in a circle, then you will have to modify these hints slightly.)

(a) To cut the chain into seven separate links only three links have to be cut. Which ones?

(b) On the first morning a link has to be cut. But which one should the woman cut? (Think carefully!)

(c) What is the longest chain for which only two links have to be cut? (It is a lot longer than you might think.) What is the longest chain for which only three links have to be cut?

Investigation

We can restrict to whole numbers of pints (or whatever), both for the given quantity of milk M, and for the capacities C, C^* of the two cans. Why is this? If we then want to divide the milk into two equal parts, the given quantity M must be even. You might begin by exploring easy cases fairly systematically in search of ideas.

M	4	6	6	8	8	8	10	10
C, C^*	1,3	1,5	2,4	1,7	2,6	3,5	1,9	
Works? Or not?						✓		

11

Circles and spheres

1. The outer circle has radius 1. Each of the six identical smaller circles touches its two neighbours and the big circle. What is the radius of the largest circle that will fit in the central hole?

2. The outer circle has radius 1. Each of the four identical smaller circles touches its two neighbours and the outer circle. What is the largest circle that will fit in the central hole?

3. The square has sides of length 4. The four identical circles fit tightly inside the square. What is the largest circle that will fit in the central hole?

4. Three identical circles fit tightly inside an equilateral triangle. What is the largest circle that will fit in the central hole?

5. Five identical circles fit tightly inside a regular pentagon. What is the largest circle that will fit in the central hole?

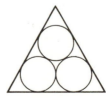

6. Six identical circles fit tightly inside a regular hexagon. What is the radius of the largest circle that will fit in the central hole?

7. Eight identical spheres fit tightly inside a cubical box. What is the largest sphere that will fit in the central hole?

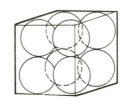

*8. (a) Four identical spheres fit tightly in the corners of a regular tetrahedron. What is the largest sphere that will fit in the central hole?

(b) Six identical spheres fit tightly inside a regular octahedron. What is the largest sphere that will fit in the central hole?

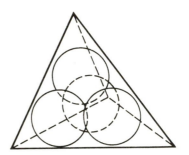

Investigation

You have a supply of discs of diameter D (e.g. pennies). What is the largest number of non-overlapping discs one can fit inside a square of side 2D? 3D? 4D? 5D? 6D? 7D? 8D?

Commentary

1. This is one of those facts that everyone should know, though it is slightly disguised here. If you are really stuck, look at one of the triangles formed by the centre of the big circle and the centres of two small touching circles. What kind of triangle is this?

2. Draw a radius of the outer circle which goes through the centre of one of the small circles.

3. Join the centre of the square to one of the corners.

4., 5. You may have to work these out in two stages. The Commentaries on Questions 1 and 3 above may help you to get started. (What matters here is the size of the central hole *relative to the rest of the figure*. Since no lengths are given, you are free to choose the size of the shape in each question. You could choose the length of each side of the triangle or the pentagon to be 1 unit, or 2 units, or 5 units. Or you may prefer to fix the radius of the small circles.)

6. You should feel that you have seen this somewhere before.

7. Join the centre of the cube to one of the corners. Which other important point lies on this line?

8. Like many 3-dimensional problems, these may appear far too hard at first sight. But they are not really all that hard, and there are several different ways of tackling them. Here is one way, which only uses very elementary geometry – together with the less familiar fact that the centre of a regular tetrahedron lies exactly one quarter of the way up each altitude.

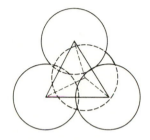

(a) If the regular tetrahedron *T* has edges of length *a*, then we want to find the size of the central hole in terms of *a*. Now suppose we start not with the regular tetrahedron *T*, but with the four identical spheres – stacked like cannonballs. We can then imagine the tetrahedron *T* fitting tightly round these spheres. We shall assume for convenience that these spheres have radius 1. Their centres form a small regular tetrahedron *T**, whose edges have length 2. It is not too hard to calculate the size of the largest sphere that will fit in the central hole. The tetrahedron *T* is an enlargement of the tetrahedron *T**. Finding the scale factor of this enlargement requires a little insight, but no fancy mathematics.

(b) This is nowhere near as hard as you might think at first sight. If you forget about the spheres that fit in the top and bottom corners for the moment, the centres of the four remaining spheres form a square.

(i) Can you see why?

(ii) Can you explain why any small sphere which fits in the central hole formed by these four spheres will still fit when we replace the spheres in the other two corners?

Investigation
Can you see how to fit sixty-eight discs into a square of side 8D?

12

Spotting factors

There are some lovely mathematical puzzles for which you need to be able to tell 'at a glance', without doing the division, when a number has certain factors. The first five questions in this section look at ways of spotting easy factors. **You should not need a calculator for any of the questions in this section**.

1. It is easy to tell at a glance, simply by looking at the last digit, whether a number is a multiple of 2. Which other factors can you spot in the same way – just by looking at the last digit of a number?

2. (a) 1234, 123456, and 12345678 are all even. Can you see how to tell 'at a glance' – well, maybe not quite 'at a glance', but without actually doing the division – which of them are divisible by 4?

(b) Test your rule by deciding 'at a glance' (without doing the division) which of these numbers are divisible by 4:

84, 558, 1234, 9876, 12354, 123456, 345678, 987654, 666626666, 82222228.

Then check each one carefully to see whether you were right.

3. (a) Find out how to tell whether a number is a multiple of 3 without doing the division. Which of the numbers in Question 2(b) are multiples of 3?

(b) Now try to explain why your rule works.

4. Can you see how to tell at a glance whether a number is a multiple of 6? Which of the numbers in Question 2(b) are multiples of 6?

5. How can the method you found in Question 3 be used to tell at a glance whether a number is a multiple of 9? Which of the numbers in Question 2(b) are multiples of 9?

6. (a) Can you find a four-digit number 1*** which uses the digits 1, 2, 3, 4 once each and which is a multiple of 4? How many different answers are there?

(b)   Can you find a number which uses the digits 1, 2, 3, 4, 5 once each, and which is a multiple of 4 but not a multiple of 6?

(c)   27*36*5 is a multiple of 225. How many different ways are there of filling in the two missing digits?

(d)   *679* is a multiple of 72. How many different ways are there of filling in the two missing digits?

(e)   Could the thirteen digit number 1234567891011 be a perfect square?

(f)   The digits of a certain ten-digit number include one 1, two 2's, three 3's, and four 4's (in some order). Could it be a perfect square?

7.   The number 123654 uses each of the digits 1–6 once each and has the interesting property that:

1	is a multiple of 1
12	is a multiple of 2
123	is a multiple of 3
1236	is a multiple of 4
12365	is a multiple of 5
123654	is a multiple of 6

(a)   Check each of these facts *without actually doing the divisions*.

(b)   Are there any nine-digit numbers which use each of the digits 1–9 exactly once each and which have a similar property? If so, how many such numbers are there?

(c)   Are there any eight-digit numbers which use each of the digits 1–8 exactly once each and which have a similar property?

8.   (a)   Pick any two-digit number; turn it back to front and subtract the smaller number from the larger. If your answer *cd* has first digit zero, treat the answer as a *two-digit* number with first digit $c = 0$. Now turn this two-digit answer *cd* back to front and add. What answer do you get? Repeat the whole procedure with several different starting numbers *ab*. What seems to happen each time? Can you explain why?

(b)   Pick any three-digit number; turn it back to front and subtract the smaller number from the larger. (If your answer *def* has first digit zero, treat the answer as a *three digit* number with first-digit $d = 0$.) Now turn this three-digit answer *def* back to front and add. What answer do you get? Repeat the whole procedure with several different starting numbers *abc*. What seems to happen each time? Can you explain?

9. (a) What digit should you put in place of the central zero in the number 3000003 to get a multiple of 11? A multiple of 13?
   (b) Choose the missing digits in 30*0*03 to get a multiple of 13.

Investigation 1
Is it possible to tell at a glance whether a number is a multiple of 11? (The most familiar multiples of 11, namely 11, 22, 33, and so on, suggest that there may be some rule. Is there a simple rule which works for all numbers, no matter how many digits they have?)

Investigation 2 (Do Investigation 1 first.)
I am thinking of a three-digit number *abc*. If I divide my number by 10, the remainder is just the last digit *c*. So if I divide each of the numbers
   abc   *bca*   *cab*
by 10 and tell you the three remainders *c*, *a*, *b*, you can immediately tell what my original number was. Suppose instead I divide each of the numbers
   abc   *bca*   *cab*
by 11, and tell you the remainder each time. Can you find an easy way of 'reconstructing' my original number?

Investigation 3
Which numbers are exactly divisible by the sum of their digits?

Commentary

1. What can you say about a number which ends in 0? What do you know about any number which ends in a 5 or a 0?

2. (a) 4 is a factor of any number which ends in two zeros (since 4 is a factor of 100). And 1234 is the same as 1200 + 34. So the important question is whether 4 is a factor of the number 34 formed by the **last two digits**.

3. You want a general rule, and an explanation of why it works. Let's look at an example. How could you tell at a glance whether 684 is a multiple of 3?
       $80 = \quad 8 \times 10 \quad\quad = 8 \times 9 + 8$
and $600 = \quad 6 \times 100 \quad\quad = 6 \times 99 + 6$
so $\quad 684 = 600 + 80 + 4 = (6 \times 99 + 8 \times 9) + 6 + 8 + 4$
Since 9 and 99 are both multiples of **3**, the number in the bracket is bound to be a multiple of 3. So the important question is whether the digit-sum 6 + 8 + 4 is itself a multiple of 3. Well, is it or isn't it?

4. To say that a number is a multiple of 6 is exactly the same as saying that it is a multiple of both 2 and 3.

5. Take another look at the Commentary on Question 3. Since 9 and 99 are both multiples of **9**, the number in the bracket is bound to be a multiple of 9. So the important question is whether the sum of the digits 6 + 8 + 4 is itself a multiple of . . . .

Extension
(a)   Can you find a way of telling at a glance whether a number written in base 9 is a multiple of 8. Or a multiple of 4? Or a multiple of 2?
(b)   Does this suggest a way of spotting certain factors in base *n*?

6.   (a)   There are two possible ways of filling in the last two digits.
(b)   What can you say about the sum of the digits of such a number? Your answer to Question 3 should then tell you one of its factors. Your number has to be a multiple of 4. Can it possibly avoid being a multiple of 6?
(c)   It should be enough to notice that 225 = 9 × 25.
(d)   It should be enough to notice that 72 = . . . × . . . .
(e), (f)   What can you say about a perfect square which is a multiple of 3?

7.   (b)   Let _ _ _ _ _ _ _ _ _ be such a number.
(i)   Which digits can you put in their correct positions straight away?
(ii)   What do you know about the digits which would have to go in the second, fourth, sixth, and eighth places? What does this tell you about the digits in the first, third, seventh, and ninth places?
(iii)   This leaves exactly two possibilities for the digit in the fourth place. What are they? And why?
(iv)   What do you know about the sum of the first three digits? What does this tell you about the sum of the next three digits? So what are the possibilities for the sixth digit?
(v)   Now look at the eighth digit.
(c)   Suppose _ _ _ _ _ _ _ _ were such a number. Then _ _ _ _ _ _ _ _ 9 would have to be a solution to part (b). Why?

Extension   For which values of the base *b* can one find a number (base *b*) which uses each of the *b* − 1 non-zero digits (base *b*) exactly once, and which has a similar property.

8.   (a)   The Commentaries on Question 3 and Question 5 reveal more than we admitted at the time. Since 9 and 99 are both multiples of 9, the number in the bracket is bound to be a multiple of 9. So the remainder you get when you divide 684 by 9 is exactly the same as the remainder you get when you divide the digit-sum 6 + 8 + 4 by 9. Hence the remainders you get when you divide *ab* and *ba* by 9 must be equal. What does this tell you about the number *cd* (= *ab* − *ba*)?
(b)   The Commentary on part (a) should tell you something significant about the number *def* (= *abc* − *cba*). And after trying a few different starting numbers you should notice something interesting about the middle digit *e*. Can you explain why *e* always has this value?

Extension   You now know what happens with any two-digit starting number *ab*, and with any three-digit starting number *abc*. What do you think will happen with a four-digit starting number *abcd*? Does it really? (Are you sure?)

9. You shouldn't need any help with these questions. But they might make you begin to wonder whether there are ways of telling when a given number is a multiple of 11, or of 13, or of 7, or . . . , without actually doing the division.

Investigation 1
The explanation in the Commentary on Question 3 depended on the fact that every power of 10 is one more than a multiple of 9. This no longer works for multiples of 11: some powers of 10, such as 10^2 or 10^4, *are* one more than a multiple of 11, but others, such as 10^3, are not. What is the closest multiple of 11 to 10^3? Try to find a rule which will predict that 121, 132, and 1221 are multiples of 11 and that 12321 is not. Then explain why your rule works.

Investigation 2
Read the Commentary on Question 8(a). Then look at the rule you found in Investigation 1 for spotting multiples of 11. Can you see how to tell at a glance, *without doing the division*, what the remainder will be when you divide a number such as 962 by 11? How does this help you here?

Investigation 3
This investigation also appeared at the end of the very first group of problems. The Commentary on page 12 should help you to begin to explore what actually happens. But you should now be in a much better position to *explain* some of the curious patterns which arise.

13

Miscellany

1. (a) Can you make six prime numbers which together use each of the nine digits 1–9 exactly once? How many different ways are there of doing this?

(b) Can you make six prime numbers which together use each of the ten digits 0–9 exactly once each? How many ways are there of doing this?

2. (a) 4 and 9 are one-digit squares. And when you put them together you get 49, which is a two-digit square. Is this the only two-digit square you can get by putting two one-digit squares together like this?

(b) How many three-digit squares is it possible to get by putting together a one-digit square and a two-digit square (or a two-digit square and a one-digit square)?

(c) How many four-digit squares is it possible to get by putting two two-digit squares together?

(d) You should know what 225 and 625 are squares of. If you suspect that 225625 is a square, can you work out in your head what it would have to be the square of?

3. Work out 1089 × 9. What do you notice? Are there any other four digit numbers for which mutliplying by 9 has the same effect?

4. Find the area of the shaded region.

5. A man wears underpants, trousers, shirt, two socks, and two shoes. In how many different ways can he get dressed?

6. (a) What is the smallest number N for which $\frac{1}{2}N$ is a perfect square and $\frac{1}{3}N$ is a perfect cube?

(b) What is the smallest number N for which $\frac{1}{2}N$ is a perfect square, $\frac{1}{3}N$ is a perfect cube, and $\frac{1}{5}N$ is a perfect fifth power?

7. The identity $\sqrt{(3969)} = 63$ uses each of the digits 3, 6, 9 exactly twice.

(a) Find a similar identity $\sqrt{(***)} = **$ which uses each of the digits 2, 4, 5, 6, 7 exactly once. Could the identity $\sqrt{(***)} = **$ use each of the digits 1, 2, 3, 4, 5 exactly once?

(b) Find an identity $\sqrt{(****)} = **$ which uses each of the digits 1, 2, 4, 5, 6, 9 exactly once. Could the identity $\sqrt{(****)} = **$ use each of the digits 1, 2, 3, 4, 5, 6 exactly once?

8. A teacher writes a number on the blackboard and asks what its factors are. One pupil says the number is divisible by 1. A second says that it is divisible by 2. A third says that it is divisible by 3, and so on right round the class.

(a) Suppose the class has thirty pupils. If only two pupils make incorrect statements, and they speak one after the other, which were the incorrect statements? And how big was the number?

(b) Suppose that precisely three pupils make incorrect statements, and that they speak one after another. How big was the class?

9. In any triangle the three medians meet in a single point. Suppose we divide each side into *four* equal parts. Do the corresponding lines always meet in three's as the dots in this diagram seem to suggest?

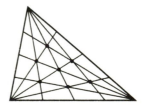

10. The list

n_0	n_1	n_2	n_3
2	0	2	0

has the curious property that n_0 ($= 2$) is exactly equal to the number of 0's in the list (since $n_1 = 0$ and $n_3 = 0$), n_1 ($= 0$) is exactly equal to the number of 1's in the list (there aren't any), n_2 ($= 2$) is exactly equal to the number of 2's in the list (since $n_0 = 2$ and $n_2 = 2$), and n_3 ($= 0$) is exactly equal to the number of 3's in the list (there aren't any). Find a list of *eight* numbers

$$n_0 \quad n_1 \quad n_2 \quad n_3 \quad n_4 \quad n_5 \quad n_6 \quad n_7$$

such that:

n_0 is equal to the number of 0's in your list
n_1 is equal to the number of 1's in your list
n_2 is equal to the number of 2's in your list

and so on.
How many different answers are there?

Investigation 1

For some numbers, like 16 or 149, the sum of the digits is a multiple of 7. Imagine all such numbers written out in one long (endless) list, smallest first. The list would start like this:

7, 16, 25, 34, 43, . . .

(a) Write down the next ten numbers in the list.
(b) The gap between successive numbers in the list varies as you go along. How large can the gap between successive numbers in the list get?

Investigation 2

(a) $2 + 2 = 2 \times 2$. Are there any other pairs of numbers a, b for which $a + b = a \times b$?
(b) Are there any triples a, b, c for which $a + b + c = a \times b \times c$? Can you find them all?
(c) What about sets of four numbers a, b, c, d with $a + b + c + d = a \times b \times c \times d$? Or sets of five numbers a, b, c, d, e? Or sets of six numbers? Or . . .

Commentary

1. (a) (1 is *not* a prime number.) Why can't 4 be the units digit of a prime number? How many of the digits 1–9 *could* be the units digit of a prime number? Write them all down. The numbers you finish up with must be prime numbers. So some of the units digits you have written down cannot possibly have a tens digit! Now look for all possible ways of arranging the remaining digits in the vacant places. (Some of the prime numbers you finish up with could have more than two digits!)

(b) You can start in exactly the same way as in part (a). The only difference is that there are more possibilities to think about at the last stage.

2. (b) There is at least one way of doing this (and possibly more than one). Start by making a list of all one-digit squares, and a list of all two-digit squares.

(c) This may look harder than part (b). But it is in fact much easier! (What could a number of the form 16** possibly be the square of? What could a number of the form 25** possibly be the square of?)

(d) It should be clear that the required number has three digits. The first and last digits are easy to find.

3. Suppose $abcd \times 9 = dcba$. If there is to be no carry to the ten thousands column, what must a be? So what must d be?

$$
\begin{array}{r}
a\ b\ c\ d \\
9\ \times \\
\hline
d\ c\ b\ a
\end{array}
$$

Extension Are there other numbers like this if we allow multipliers other than 9?

5. The answer is surprisingly large, so you will have to find a clever way of counting. (The seven items of clothing fall naturally into three two's – each of which has to be put on in a fixed order – and one item which can be put on at any stage.)

6. (a) N must be even. (Why?) So $\frac{1}{3}N$ must be even; and since $\frac{1}{3}N$ is a cube it must be a multiple of __ . N is a multiple of 3. (Why?) So $\frac{1}{3}N$ must be a multiple of 3; and since $\frac{1}{2}N$ is a square, it must be a multiple of __ . Hence $\frac{1}{3}N$ must be a multiple of 3. If you keep on like this, you will soon find the smallest possible N.

(b) This is like part (a), but a bit more complicated.

7. (a) If $\sqrt{(abc)} = de$, then c must be the units digit of a *square*, namely $(de)^2$. This leaves very few possibilities for the pair e, c.

8. (a) (i) Suppose the *ninth* statement was incorrect. Then two other statements would have to be incorrect as well. Can you say which two?

(ii) Could the eighth statement have been incorrect? (Use the same reasoning as in (i) above.)

(iii) There is no way of knowing the precise number the teacher actually wrote on the board. But can you find the smallest number which would have worked? (You might also try to find the next smallest number which would have worked.)

(b) (i) When you identified the two incorrect statements in part (a), did you use the fact that the class had exactly thirty pupils? Would you have got a different answer if the class had had seventeen pupils? Or twenty-two pupils? Or thirty-one pupils? What if the class had had thirty-two pupils?

(ii) There is no way of knowing the exact size of the class in part (b). But you can find all the possibilities for the three incorrect statements. Each of these possibilities for the three incorrect statements corresponds to a range of possible class sizes. (The smallest trio of incorrect statements is the second, the third, and the fourth, which would occur in a class of size __ or __ only. Why? What is the next smallest possible trio of incorrect statements? And the next? And the one after that?)

9. What happens in an equilateral triangle? And why does this help?

10. Don't be put off by the curious way this question is worded. Write down a list (any old list!) of eight numbers

n_0	n_1	n_2	n_3	n_4	n_5	n_6	n_7

and look at it to see whether n_0 really is equal to the number of 0's in your list, then whether n_1 is equal to the number of 1's in your list, and so on. Then try to improve your list. You should eventually stumble on a list of the kind you are looking for! You can then try to be more systematic and to find all lists with the required property.

Investigation 1
The numbers up to 100 suggest that the gap between successive numbers in the list is always either . . . or . . . . Where does this first guess break down? What happens for numbers in the list between 106 and 194? And what happens straight after 194? You may be tempted to jump to another conclusion. But you've been wrong before! Where might be a good place to look for an even larger gap? Or a smaller gap than that between 59 and __?

Investigation 2
(b) How many of the numbers a, b, c have to be 1's?
(c) How many of the numbers a, b, c, d have to be 1's?

14

Beginnings and ends

1. If you start with the number 2 and stick an extra digit 1 at the beginning you get 12, which is exactly six times the number you started with.

    (a)   Can you find a number so that when an extra digit 1 is written at the beginning you get three times the number you started with?

    (b)   Find a number so that when an extra digit 1 is written at the beginning you get five times the number you started with.

    (c)   Find a number so that when an extra digit 1 is written at the beginning you get nine times the number you started with.

    (d)   Is 2 the only number which gets multiplied by six when you stick an extra digit 1 at the beginning? And what about your answers to parts (a), (b), and (c)? Is there only one possible answer to each question?

2.   (a)   Can you find a whole number which starts with a 1, and such that when this digit 1 is moved to the other end you get three times the number you started with? Is your answer the only possible one?

    (b)   Can you find a number which starts with a 2 and such that when this digit 2 is moved to the other end you get three times the number you started with?

3.   (a)   Find a number which ends with a 3 and such that when this digit is rubbed out and written at the beginning you get three times the number you started with.

    (b)   Find a number such that when its first digit is rubbed out and written at the end you get half the original number. Is your answer the only possible one?

4.   (a)   Find a number such that when an extra digit 1 is written at each end the resulting number is ninety-nine times the number you started with.

    (b)   Find a number such that when an extra digit 2 is written at each end the resulting number is one hundred and ninety-nine times the number you started with.

(c)   Suppose we write an extra 7 at the beginning and an extra 2 at the end. Can the resulting number be 201 times the number we started with?

5.   Find a whole number such that when one of its digits is rubbed out, the resulting number is equal to one ninth of the original number, and is itself a multiple of 9. How many numbers are there with this property?

Investigation 1 (Do Question 1 first.)
Sticking an extra 1 at the beginning of a number can give you three times the number you started with – or five times, or six times, or nine times the number you started with. What's so special about *three*, *five*, *six*, and *nine*? Can you get twice, four times, seven times, or eight times the number you started with in the same way?

Investigation 2 (Do Question 2 first)
(a)   A number starts with a 1. Shifting this digit 1 to the other end gives an exact multiple of the number you started with. Which multiples are possible?
(b)   A number starts with a 1. Shifting this digit 1 to the other end produces an exact factor of the number you started with. What can you say about the possible values of the ratio of the two numbers.

Commentary

1.   (a)   Trial and error should see you through here.
   (b), (c)   Intelligent guesswork *may* be enough for parts (b) and (c) as well. It all depends how well you know your numbers! Whether or not guesswork is enough for these easy questions, you will eventually have to do some sort of calculation to find the numbers you want. There are lots of ways of doing this. One way is to think of a number you are looking for as a long 'squiggle' ∿∿∿, though you don't yet know how many digits it will have, or what they will be. In part (b), for example, you know that

   $(1\,∿∿∿) = 5 \times (∿∿∿)$.

In other words

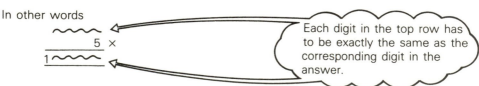

   (d)   If decimals are allowed, then 0.2 would be a different answer. If decimals aren't allowed, you could try multiplying by 10 (instead of dividing by 10), or by 100, or . . . . The kind of calculation described above for part (b) should tell you whether these are the only answers.

2. (a)   There are several different ways of tackling all of these problems. One way is the method suggested above for Question 1(b). First write the number you start with as '1 ~~~~~'. When you move the 1 to the other end, you want to land up with three times this number. So you can work out what the squiggle must be from

$$
\begin{array}{r}
1 \;\text{~~~~~} \\[-2pt]
3 \;\times \\
\hline
\text{~~~~~}\; 1
\end{array}
$$

(There are in fact infinitely many different answers. But they are all related to each other. Can you see how?)

(b)   The approach suggested above for part (a) should work here too. But it is well worth looking for a completely different method.

3. (a)   This time your starting number is '~~~~~3'. You could try multiplying this by 3 exactly as in Question 2(a). Or you could take the number you finish with, namely '3 ~~~~~', and divide by 3 to find the number you started with

$$
\begin{array}{r}
\text{~~~~~}\,3 \\
3\,\overline{)\,3\;\text{~~~~~}}
\end{array}
$$

(b)   As you might expect, the answer depends on the first digit of the original number. How many possibilities are there for the first digit? Do they all 'work'? How are the answers you get with different first digits related?

4.   These are a bit harder. But the same methods should still work. When you have done one or two you might like to look for another way of solving equations like:
$$99 \times (\text{~~~~~}) = 1\;\text{~~~~~}\;1.$$

5. (i)   Removing a digit from a number *divides the portion on its left by 10*, and leaves the portion on its right unchanged. If the overall effect of this is to divide the whole number by 9, then there is only one possible position for the digit that gets rubbed out. Where must it be? And why?

(ii)   When that digit is rubbed out, the resulting number has to be both one ninth of the original *and* an exact multiple of 9. This should tell you the value of the digit which has been rubbed out.

(iii)   As in earlier questions, there are lots of different answers. But they are all very closely related!

15

Counting: one

1. (a)   Each amoeba divides into two separate amoebas once every hour. If we start with just one amoeba, how many will there be at the end of one hour? At the end of two hours? Three hours? And so on. (Mathematical amoebas never die!)

Time	Start	1 hour	2 hours	3 hours	4 hours	5 hours
Number of amoebas	1					

(b)   Mathematical rabbits come in breeding pairs. They start to breed at the age of two months. They then produce one breeding pair of rabbits every month. If we start with just one newly born pair of rabbits, how many pairs will there be at the end of one month? At the end of two months? Three months? And so on. (Mathematical rabbits never die!)

Time	Start	1 month	2 months	3 months	4 months	5 months
Number of breeding pairs	1					

2.   The number of amoebas at the start, after 1 hour, after 2 hours, and so on, is just the familiar sequence of powers of two.

(a)   Write down the first ten powers of two on one line. Write down the difference between the second number and the first number, the difference between the third number and the second number, the difference between the fourth number and the third number, and so on. What do you notice? Can you explain?

(b)   Write down the first ten powers of two on one line. Write the difference between the third number and the first number below the second number. Write the difference between the fourth number and the second number below the third number, the difference between the fifth number and the third number below the fourth number, and so on. What do you notice? Can you explain why this happens?

(c)   Write the first ten powers of two again on one line. Multiply the first and third numbers, and write the answer below the second number. Multiply the second and fourth numbers and write the answer below the third number. Keep going. What do you notice? Can you explain?

(d)   Write down the first ten powers of two again. Write the first number below the second number. Write the sum of the first two numbers below the third number. Write the sum of the first three numbers below the fourth number. Keep going. What do you notice? Can you explain?

3.   In Question 1(b), the number of pairs of rabbits at the start, after 1 month, after 2 months, after 3 months, and so on, is called the sequence of *Fibonacci numbers*.

(a)   Write down the first ten Fibonacci numbers on one line. Write down the difference between the third number and the second number, the difference between the fourth number and the third number, the difference between the fifth number and the fourth number, and so on. What do you notice? Can you explain?

(b)   Write the first ten Fibonacci numbers on one line. Write the difference between the third and first numbers below the third number. Write the difference between the fifth and third numbers below the fourth number. Keep going. What do you notice? Can you explain?

(c)   Write down the first ten Fibonacci numbers again. Multiply the first and third numbers, and write the answer below the second number. Multiply the second and fourth numbers, and write the answer below the third number. Keep going. What do you notice? Will this go on happening? Explain.

(d)   Write down the first ten Fibonacci numbers once more. Write the first number below the *third*. Write the sum of the first two numbers below the *fourth*. Write the sum of the first three numbers below the fifth. Keep going. What do you notice? Can you explain why this happens?

4.   Here is the beginning of a pattern of numbers called *Pascal's triangle*. It has 1's all the way down each of the two sloping sides. Each number in the middle is got by adding the two numbers directly above it.

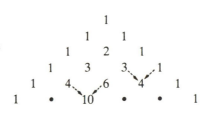

(a)   Make a Pascal's triangle of your own. Fill in the numbers in the first eleven rows. (The number in the middle of the eleventh row should be 252.)

(b) Add up the numbers in each row. What do you get? Can you explain why the sum of all the numbers in each row is always *exactly* *double* the sum of all the numbers in the previous row?

(c) Suppose you combine the numbers in each row by alternately adding and subtracting (for example, $1 - 4 + 6 - 4 + 1$). What do you get? Can you explain?

(d) Add up the numbers along each of these sloping lines. What do you get? Can you explain why this happens?

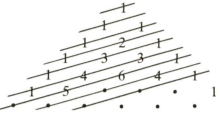

5. (a) Which Fibonacci numbers are even? (Is the 998th Fibonacci number even? What about the 999th? And what about the 1000th Fibonacci number?) Can you explain why your rule works?

(b) Which Fibonacci numbers have 3 as a factor? (Is the 998th Fibonacci number a multiple of 3? What about the 999th? And the 1000th?) Can you explain why your rule works?

(c) Which Fibonacci numbers have 4 as a factor? Which have 5 as a factor? Which have 6 as a factor?

6. (a) For each Fibonacci number make a square whose side has that Fibonacci number as its length. These squares (1 by 1, 1 by 1, 2 by 2, 3 by 3, 5 by 5, 8 by 8, and so on) fit together exactly as shown here. The first 1 by 1 square has been shaded. Mark the centres of each of your squares as accurately as you can. Do these centres conform to any recognisable pattern? If you think they do, can you prove that they do?

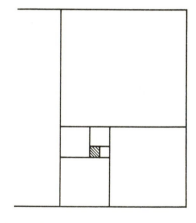

(b) Write down the first ten Fibonacci numbers. Write the square of the first number below the second number. Write the sum of the squares of the first two numbers below the third number. Write the sum of the squares of the first three numbers below the fourth number. And so on. What do you notice? Can you explain?

7. Each number in the Fibonacci sequence
$$1, 1, 2, 3, 5, 8, 13, 21, 34, 55, 89, 144, \ldots .$$
is equal to the sum of the two previous numbers in the sequence.

(a) Work out the ratios of successive terms $\frac{1}{1}, \frac{2}{1}, \frac{3}{2}, \frac{5}{3}, \frac{8}{5}$, and so on. What do you notice?

(b) Now plot the points with coordinates (1,1), (2,1), (3,2), (5,3), (8,5), and so on, on squared paper. What do you notice?

8. (a) The sequence
$$a, b, a + b, \ldots .$$
starts with the two unknown numbers a and b. Each of the other numbers in the sequence is equal to the sum of the two previous numbers – just like the Fibonacci sequence. Work out the next ten numbers in the sequence. What do you notice? Can you explain?

(b) Given the two lengths a and b, work out the lengths c, d, e, f, etc.

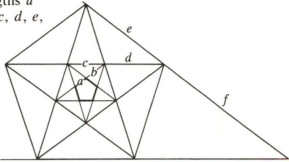

Investigation 1 (Do Question 4 first.) This pattern is called the *harmonic triangle*. Write down the numbers in the next row. In what ways is the harmonic triangle related to Pascal's triangle?

$$1$$
$$\frac{1}{2} \quad \frac{1}{2}$$
$$\frac{1}{3} \quad \frac{1}{6} \quad \frac{1}{3}$$
$$\frac{1}{4} \quad \frac{1}{12} \quad \frac{1}{12} \quad \frac{1}{4}$$
$$\frac{1}{5} \quad \frac{1}{20} \quad \frac{1}{30} \quad \frac{1}{20} \quad \frac{1}{5}$$
$$\frac{1}{6} \quad \frac{1}{30} \quad \frac{1}{60} \quad \frac{1}{60} \quad \frac{1}{30} \quad \frac{1}{6}$$

Investigation 2
(a) Every number can be written in several ways as a sum of 1's and 2's. For example, $3 = 1 + 2$ and $3 = 1 + 1 + 1$. In how many ways can the number 11 be written as a sum of 1's and 2's? In how many ways can the number 73 be written as a sum of 1's and 2's? Find a general rule and explain why it works.
(b) We would not usually treat $3 = 1 + 2$ and $3 = 2 + 1$ as different. But if we do, then there are *three* different ways of writing 3 as a sum of 1's and 2's. What are they? And if we count in the same way, how many different ways are there of writing 11 as a sum of 1's and 2's? Investigate!

Investigation 3
How many odd numbers are there in the nth row of Pascal's triangle?

Commentary

1. (a) By the end of one hour the first amoeba will have divided. By the end of the second hour both amoebas will have divided again. And so on.

(b) At the end of the first month the starting pair will still be too young to breed. But by the end of the second month they will have produced their first pair of offspring. By the end of the third month the original pair will have given birth to a second breeding pair, though their first pair of offspring will still be too young to breed and will not start breeding until the fourth month.

2. (a) The differences are themselves powers of 2. Why is this?

(b) The differences (3, 6, 12, . . .) are closely related to the corresponding powers of 2 (2, 4, 8, . . .). What is the connection?

(c) Write each product in the second row as a power of 2.

(d) (i) Look what happens, for example, if you add 1 to $1 + 2 + 2^2 + 2^3$:

$$\underline{1 + (1} + 2 + 2^2 + 2^3) = \underline{2 + 2} + 2^2 + 2^3$$
$$= \underline{2^2 + 2^2} + 2^3$$
$$= \underline{2^3 + 2^3}$$
$$= \underline{2^4}$$

(ii) Factorize $(x^4 - 1) = (x - 1)(x^3 + \ldots\ldots)$. What does this tell you when $x = 2$?

3. (a) Compare $2 - 1 = 1$ and 1; compare $3 - 1 = 2$ and 2; compare $5 - 2 = 3$ and 3; . . . . Each Fibonacci number is just the *sum* of the two previous Fibonacci numbers. How does this help to explain?

(b) In Question 2(b) each difference was exactly $1\frac{1}{2}$ times the corresponding power of 2. This time the relationship is even simpler! Why?

(c) (i) In Question 2(c) each product was *equal* to the square of the corresponding power of 2. This time the relationship is not quite so simple. But you should still be able to see what it is, though you will probably need to use some algebra to explain *why* it happens.

(ii) Let $F_1, F_2, F_3, F_4, \ldots$ stand for the sequence of Fibonacci numbers; so:

$$F_1 = 1, F_2 = 1, F_3 = F_1 + F_2, F_4 = F_2 + F_3, F_5 = F_3 + F_4, \text{ and so on.}$$

Use $F_3 = F_1 + F_2$ and $F_1^2 - F_2^2 = 0$ to work out $F_3F_1 - F_2^2$.

(iii) Now look at $F_4F_2 - F_3^2$. **Do not substitute values.** Instead use the equations $F_4 = F_2 + F_3$ and $F_3 = F_1 + F_2$ (so $(F_2 - F_3) = -F_1$) to show that $F_4F_2 - F_3^2 = -(F_3F_1 - F_2^2)$.

(iv) Then use the same trick to show that $F_5F_3 - F_4^2 = -(F_4F_2 - F_3^2)$. Will the same trick always work? Why?

(d) To explain why this happens you will probably have to use algebra and avoid substituting actual values (as in part (c)).

(i) Use $F_3 = F_1 + F_2$ to work out $F_3 - F_1$.

(ii) Use $F_4 = F_2 + F_3$ to show that $F_4 - (F_1 + F_2) = F_3 - F_1$.

(iii) Use $F_5 = F_3 + F_4$ to show that $F_5 - (F_1 + F_2 + F_3) = F_4 - (F_1 + F_2)$.

(iv) Will the same trick always work?

4. (b) (i) Look at the way each number in the fifth row is equal to the sum of the two numbers directly above it in the fourth row. How many times does each number in the fourth row get used when working out the five numbers in the fifth row?

(ii) Think how you would work out the fiftieth row from the forty-ninth row. Does the same reasoning apply?

(c) There is only one number in each of the first two sloping lines, namely 1. So the sequence of numbers you get certainly begins 1, 1, ... If you can show that the sum of the numbers in each sloping line is *always* equal to the sum of the numbers in the two previous sloping lines, then you can be sure that the sequence of numbers will continue:

1, 1, 2, 3, 5, 8, 13, 21, ...

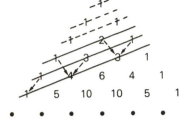

(i) Can you see why the sixth sloping line (1 + 4 + 3) has to be equal to the sum of the numbers in the two previous sloping lines?

(ii) Can you see why the seventh sloping line (1 + 5 + 6 + 1) has to be equal to the sum of the numbers in the two previous sloping lines?

(iii) In what way is every even-numbered sloping line like the sixth, and every odd-numbered sloping line like the seventh?

5. (a) Each Fibonacci number is equal to the sum of the two previous Fibonacci numbers. Suppose you knew that the first two Fibonacci numbers were both odd, but you did not know their actual values. What would this tell you about the third Fibonacci number? And the fourth? And the fifth? And the sixth? Why must this pattern go on forever?

(b) A number is *even* if it leaves remainder 0 when you divide by 2; a number is *odd* if it leaves remainder 1 when you divide by 2.

(i) How many possible remainders are there when you divide by 3? What are they?

(ii) Each Fibonacci number is equal to the sum of the previous two. Suppose you knew that the first two Fibonacci numbers both leave remainder 1 when you divide by 3, but you did not know their actual values. What would this tell you about the remainder you would get when you divide the third Fibonacci number by 3? And the fourth? And the fifth? And the sixth? And the seventh? And the eighth?

6. (a) Look at the centres of the first square, the third square, the fifth square, the seventh square, and so on. Then look at the centres of the other squares. Try to explain what you find.

(b) You should certainly notice *something*. To explain why this happens, look for a connection with part (a).

7. (a) Try to make your observation as specific as you can. The ratios seem to be getting closer to something. But what number are they getting closer to? And are they creeping *up* towards the number from below? Or *down* towards the number from above? Or what?

(b) Remember how the ratios behaved in part (a).

8. (b) The inner pentagon is regular, so AC must be parallel to ED, and BE must be parallel to CD. Thus XCDE is a _____ .

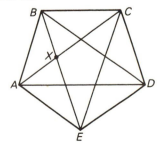

Investigation 1

It should be clear how the next row *begins*. But what is the second entry? Look back at the previous two rows and try to see why the entry after $\frac{1}{6}$ has to be $\frac{1}{30}$. (What is the connection between $\frac{1}{6}$, $\frac{1}{30}$, and $\frac{1}{5}$?)

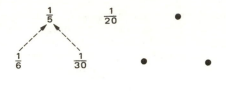

Once you think you can see how to build up the harmonic triangle, there are lots of interesting questions to explore. What can you say about the sum of the entries in each row? What about alternately adding and subtracting the numbers in each row? Is there any connection between successive rows? Is there any connection between the *n*th row of the harmonic triangle and the *n*th row of Pascal's triangle?

16

Polygons

1. What is the ratio of the areas of the inner and outer regular hexagons in this figure?

2. A polygon inscribed in a circle has all its angles equal, but its sides are not all equal. Show that it must have an even number of sides.

3. *ABCDE* is a pentagon with all five sides equal in length.
   (a) Can you be sure that all five angles are equal?
   (b) Suppose the five angles at *A, B, C, D, E* satisfy $A \geqslant B \geqslant C \geqslant D \geqslant E$. Must the pentagon be regular?

4. Is the large regular pentagon more than twice, or less than twice, the area of the star inside it?

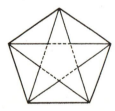

5. (a) What is the largest equilateral triangle that can be inscribed in a regular hexagon?
   (b) What is the largest regular hexagon that can be inscribed in an equilateral triangle?

6. (a) What is the largest equilateral triangle that can be inscribed in a square? (When you think you know, try to explain why you think it is the largest.)
   (b) What is the largest equilateral triangle that can be inscribed in a regular pentagon?

7. (a) What is the largest square that can be inscribed in an equilateral triangle? (When you think you know, try to explain why you think it is the largest.)

(b) What is the largest square that can be inscribed in a regular hexagon?

Investigation 1

(a) Mark five points on a circle. Make sure they are *not* equally spaced. Join each point to the next but one round the circle in each direction. What is the sum of the angles at the five points of this 5-pointed star?

(b) What angle-sum do you get for a 7-pointed star in which each point is joined to the next but one round the circle? What if each point is joined to the next but two round the circle?

Investigation 2

Here is a 'closed' path with five straight edges, which crosses itself in the three points A, B, C. How often can the edges of such a closed path cross?

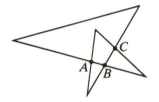

Commentary

1. (i) It is not necessary to work out areas. If you do, then perhaps you should ask: what is the ratio of the areas of the two triangles *CBD* and *ABC*? (Each triangle is exactly _____ of the corresponding hexagon.)

(ii) You can avoid working out areas once you realise that the areas of similar shapes are proportional, and that the ratio of their areas is equal to the ratio of the *squares* of any length you choose. So choose corresponding 'lengths' in the two hexagons whose squares are easy to compare.

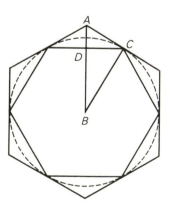

2. If the sides are not all equal, then there must be two adjacent sides, say *AB* and *BC*, that are unequal. If all the angles have to be equal, what can you say about the length of the next side *CD*?

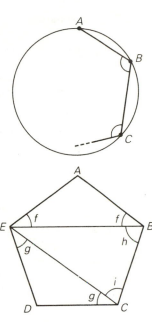

3. (a) If a quadrilateral *ABCD* has all four sides equal in length, can you be sure that all four angles are equal? (Careful!)

(b) Suppose *A, B, C, D* are not all equal. Then *A* must be greater than *D*. What does this tell you about *f* and *g*? You are also told that *B* ⩽ *C*. What does this tell you about *h* and *i*? And what would this imply about the two sides *EB* and *EC* of the triangle *EBC*? So . . . .

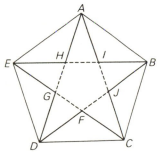

4. (i) You could try to do this by calculating each area approximately (using a calculator) and then comparing the two answers. But you should want to find a 'more mathematical' solution than that.

(ii) One approach is to compare each of the five extra outside pieces, like *ABI*, with one fifth of the inside star, or one half of *ABI* with one tenth of the inside star. Is there a natural way of cutting up the star into five, or ten, equal pieces? Can you find a good way of comparing one of the five pieces with the triangle *ABI*? Or one of the ten equal pieces with half the triangle *ABI*?

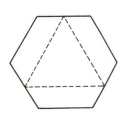

5. It should not be too hard to *guess* the correct answers to these questions. (Are you sure your guesses are larger than the ones shown here?) The real difficulty in this question, and the next two, lies in proving beyond all shadow of doubt that your guesses are correct. In this particular question you may feel so sure that you are right that any further proof seems unnecessary. But in some of the later questions it may not be so easy to guess correctly; and it may be even harder to explain why your guess really is the largest. So you should practise looking for reasons why in these two relatively easy problems.

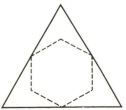

6. (a)   (i)   Could it be the one shown here?

    (ii)   Suppose you rotate this triangle slightly about the bottom right hand corner. Couldn't you then enlarge it so that it just touches the left hand edge of the square?

    (iii)   Continue turning about the bottom right hand corner, enlarging as you go. When do you have to stop? Why?

    (iv)   Can you be absolutely sure that no other equilateral triangle inscribed in the square could possibly be bigger than the one you have found? Try to explain.

    (b)   The same ideas as those in the Commentary on part (a) (i)–(iii) should lead you to make a convincing guess. Proving that your guess really is the largest may not be so easy.

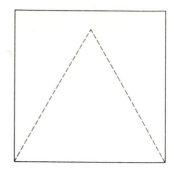

7. (a)   (i)   Is it possible to fit a square into an equilateral triangle like this?

    (ii)   Which of the first two inscribed squares shown here is the larger? Explain.

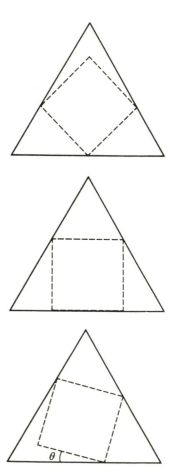

    (iii)   How do the first two inscribed squares compare with the largest inscribed square inclined at angle θ to the base? How does the area of the inscribed square vary as θ increases from $0°$ to $60°$?

17

Counting: two

1. (a)   Five points are marked on the circumference of a circle. How many chords can you draw joining these five points?

(b)   How many chords could you draw if you started with ten points round the circle? What if you started with fifty points round the circle?

2. (a)   You have a row of ten discs, five dark and five light. They alternate:

dark, light, dark, light, dark, . . . from left to right. You want to get all the dark discs to the right hand end, and all the light discs to the left hand end. The only moves you are allowed to make are those which interchange the positions of two neighbouring discs. How many moves does it take?

(b)   How many moves would it take if you started with ten dark and ten light discs? What if you started with fifty dark and fifty light discs?

3. (a)   Which of these two numbers is bigger:
$$1986 \times (1 + 2 + 3 + 4 + 5 + \ldots + 1987) \text{ or}$$
$$1987 \times (1 + 2 + 3 + 4 + 5 + \ldots + 1986)?$$

(b)   Which of these two numbers is bigger:
$$100^2 - 99^2 + 98^2 - 97^2 + 96^2 - \ldots - 1^2 \text{ or}$$
$$100 + 99 + 98 + 97 + 96 + \ldots + 1?$$

4. (a) One way of sorting a list of names into alphabetical order is to start at the beginning of the list and interchange neighbouring pairs which are obviously in the wrong order. What is the largest number of interchanges one would ever need when sorting a list of five names?

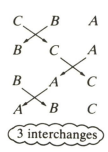

3 interchanges

(b) What if the list had ten names? Or fifty names?

5. Two straight lines cut the plane into either three regions (if the lines are parallel), or four regions (if the lines cross).

(a) What is the *smallest* number of regions you can get with three lines? With five lines? With ten lines? With fifty lines?

(b) What is the *largest* number of regions you can get with three lines? With five lines? With fifty lines?

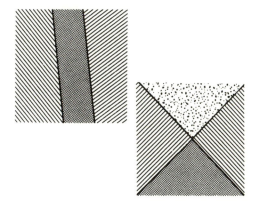

6. (a) Three boys and three girls face each other on a row of seven stools – with the boys on the left, the girls on the right, and an empty stool in between. You want to switch the boys to the right-hand end, and the girls to the left-hand end. Boys and girls can only move 'forwards' – boys to the right, and girls to the left – either by moving one place directly onto a vacant stool, or by jumping over one member of the opposite sex onto a vacant stool. Is it possible to switch the positions of the boys and the girls in this way? If so, how many moves does it take?

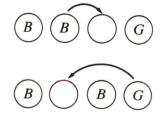

(b) How many moves would it take if you started with five boys and five girls (and one empty stool)? Or ten boys and ten girls? Or fifty boys and fifty girls?

Investigation 1 (Do Question 5 first.)
In Question 5 you found the smallest and the largest number of regions one can get with three lines (or five lines, or fifty lines). Is it always possible to position the (three, or five, or fifty) lines to get any number of regions between the smallest number and the largest number?

Investigation 2
(a) *ABCDEFG* is a polygon with seven sides. Some triangles, such as *ABC* or *ADE*, make use of *sides* of the polygon. Others, such as *ACF*, use only *diagonals*. How many triangles are there which use only diagonals?
(b) What happens for polygons with different numbers of sides?

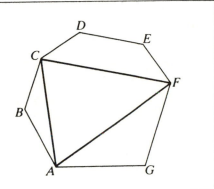

Commentary

1. (b) It is not hard to answer the question for five points. But the whole thrust of this problem is to discover how to answer the question for *any number* of points – whether fifty, or five hundred, or . . . . If you think you should be able to answer this general question, try to do so on your own. When you get stuck, read on!
(i) It may be a good idea to start by exploring the very simplest questions of this kind: 'How many chords are there if I have just one point on the circle? What if I have two points? Three points? . . .' and so on. Make a table like this.

Number of points on the circle: N	1	2	3	4	5	6	7	.	.	.
Number of chords: C	0	1								

How do you think it goes on? Guess the next few entries. Then check your guess by counting. Explain why the number of chords goes up in steps in this way.
(ii) Extend the table to finding the number of chords you would get if you started with ten points on the circle.
(iii) Extending the table is not a very efficient way of finding the number of chords when there are fifty, or five hundred, points on the circle. Can you find a quick and direct method for calculating the number of chords, C, from the number of points, N? Any method will do, but a **formula** would be most useful.

2. (b) It is easy to count the number of moves when you start with five dark and five light discs (though you may not have managed to explain yet why it cannot possibly be done in fewer moves). But the question is really concerned with discovering what happens in general, when you start with an arbitrary number of discs: you would then be able to deduce what happens when you start with fifty, or fifty thousand, discs of each colour. If you can see how to tackle this on your own, have a go. If you can't, it might be a good idea to explore the problem systematically in the hope of 'spotting' some kind of pattern, which you could then try to explain.

(i) Make a table and fill in the first few entries *carefully*.

Number of dark discs: N	0	1	2	3	4	5	6	7	.	.	.
Number of moves: M	0	1									

(ii) Our brains always try to see a pattern – even when there is none to see! (Most optical illusions, like the one on the cover, arise from our determination to force new images into familiar patterns.) So if you think you can see a pattern, you had better try to explain why the table has to carry on in the way you expect it to.

(iii) If you want to know the number of moves required when we start with fifty, or five hundred, dark discs, extending the table is not a very efficient way of working it out. You need a **formula** which tells you the number of moves, M as soon as you know the number of dark discs, N.

3. (a) The formula you should have found in Question 1(b) would obviously help here. If you have not yet managed to find a formula, you should still be able to get started by expanding the right hand side like this:

$1987 \times (1 + 2 + 3 + 4 + 5 + \ldots + 1986)$
$$= (1986 + 1)(1 + 2 + 3 + 4 + 5 + \ldots + 1986)$$

(b) You should know how to simplify the difference of two squares. What does this tell you about $100^2 - 99^2$? And $98^2 - 97^2$? . . .

4. (a) It may not be obvious how to do this at first. But it is not hard once you have the right idea. The largest number of interchanges will be needed when the list is most out of order. It is not unreasonable to guess that a list is most out of order when it is completely back to front. *But this is only a guess.*

(i) When sorting a list by interchanges, would it ever be necessary to interchange the same pair of names twice over?

(ii) So what is the largest imaginable number of interchanges one would ever have to make when sorting a list of five names (say A, B, C, D, E) in this way? (How many different pairs are there?) Can you find an arrangement of five names which requires exactly this number of interchanges?

(b) Now apply the same ideas to lists of ten, or fifty, names.

5. (a) You have to imagine the straight lines going on forever in both directions. To get the *smallest possible* number of regions with three lines (or five lines, or ten lines) you have to think how to position the lines relative to each other. One line always creates two regions. Each new line cuts at least one region in half. So the smallest number of regions you can possibly get with five lines is ___ ; and with fifty lines is ___ .

(b) To get the largest possible number of regions you want each new line to cut as many old regions as possible.

(i) Make a table.

Number of straight lines: N	0	1	2	3	4	5	6	7	.	.	.
Largest possible number of regions: R	1	2	4	.							

(ii) Don't be misled by the first three values of R. If you fail to see a pattern emerging, check your values of R for three lines, four lines, and five lines. Then look at the way the value of R increases.

(iii)   If you are still not sure what is going on, add a third row to your table, listing the value of $R - 1$ underneath each value of R.

(iv)   Now try to explain *why* the largest possible number of regions goes up in steps like this. (What is the largest number of new regions you could possibly create when you add a third line? A fourth line? A fifth line? And so on.) Then find a formula which expresses R in terms of N.

Extensions

(a)   Two planes cut 3-dimensional space into either 3 regions (if they are parallel), or 4 regions (if they cross). What is the largest number of regions you can get with three planes? With five planes? With ten planes?

(b)   If you mark three points on a circle and draw all the chords between them, then you cut the circle into 4 regions. What is the largest number of regions you can get with four points round a circle? With five points? With ten points?

(c)   If you draw one circle in the plane, it cuts the plane into two pieces. What is the largest number of regions you can create with two circles? With three circles? With four circles? With five circles? With ten circles?

6.   (a)   Think carefully at each stage whether the move you are about to make is likely to mess things up in one or two moves' time.

(b)   You should not have too much difficulty either with three boys and three girls, or with five boys and five girls. But some general insight is needed before you are likely to discover how many moves will be required when you start with *fifty* boys and *fifty* girls. As before, a table may help you to guess a pattern. But you will need more than a table to explain *why* the number of moves is always what you think it is. If you fail to find your own explanation, the following observations may help.

(i)   Suppose you start with three boys (on the left) and three girls (on the right), and move the right-hand boy first. You will eventually reach a stage where the boys and girls alternate with an empty stool on the left: _ B G B G B G. How many moves does it take to reach this stage? How many moves would it take to reach the corresponding stage if you started with one boy and one girl? With two boys and two girls? With four boys and four girls?

(ii)   You have no choice about what to do next. How many extra moves does it take to get the girls and boys alternating with the empty stool on the right: G B G B G B __ ? How many extra moves would be required to get to this stage if there were five boys and five girls?

(iii)   Your position is now the mirror image of the position you reached in (i). So you should be able to predict how many moves will be needed to reach your goal G G G _ B B B (which is the mirror image of the position you started from).

Extension 1   If three girls and four boys face each other on a row of eight stools with an empty stool between them, how many moves does it take for them to swap ends? How many moves will it take with seven girls and ten boys? Can you get a rule to cover all cases?

Extension 2   What if the boys and girls start out with more than one empty stool between them? How many moves will it take if there are B boys, G girls, and E empty stools between them?

Investigation 1

You may find that the question posed has a very simple answer. But you should not stop there. Which numbers of regions is it possible to get with three lines? With four lines? With N lines?

Investigation 2

(b)   An important part of the exercise here is to find ways of counting which are efficient *and* accurate. For polygons with less than six sides the answer is very simple. You should count the number of triangles of the required kind in polygons with 6, 7, 8, 9, 10 sides. Even then you should not expect it to be easy to see what is going on. Have a really good go at making sense of the numbers you get before reading on.

There are lots of different ways of trying to make sense of these numbers. Here is one way of beginning (though it still leaves you with rather a lot of explaining to do at the end). Suppose that instead of counting ordinary triangles of the required kind in a polygon with n sides we count clockwise 'ordered' triangles: then each ordinary triangle corresponds to three different 'ordered' triangles, *XYZ, YZX, ZXY*, so we will get three times as many 'ordered' triangles as ordinary triangles. Moreover, each time we find *one* ordered triangle *XYZ* we get n different ordered triangles just like it by 'rotating' the triangle *XYZ* round the polygon. So ordered triangles come in groups of n. All this suggests that if we multiply the number of *ordinary* triangles of the required type by 3 and divide by n, we will get the number of different kinds of 'ordered' triangles. Try multiplying the numbers of ordinary triangles you found by $\frac{3}{n}$ for $n = 6$, 7, 8, 9, 10. What do you get? Does this help?

18

Endless sums

You know how to do ordinary addition. Well, here are some extraordinary additions. The numbers are all in columns, so you can add them up in columns, 'carrying' to the left as you normally would. But there is no right hand column! So you will have to start at the *left* hand end. But as you go along you will have to look ahead to see whether the next column or two are likely to produce a 'carry'.

1. (a)   The whole numbers go on forever. In this sum the units digit of each number moves over one column further to the right. The dots at the end mean that the sum is supposed to go on for ever. But can you work out what the answer is going to be?

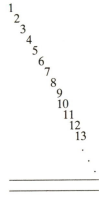

   (b)   Here is another endless sum. This time we start with a 3, and go up in steps of 3. As before, the units digit of each number moves over one column further to the right; and the dots at the end mean that the sum goes on forever. Can you work out what the answer is going to be?

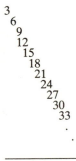

    (c) What is the connection between the sums in part (a) and (b)? How could you have got the answer to (b) straight from your answer to (a)?

2. (a) Can you work out the answers to these two?

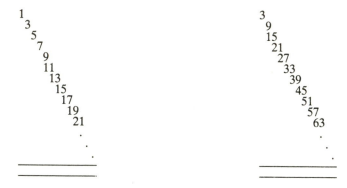

    (b) What is the connection between your two answers? How could you have got the answer to the second one straight from your answer to the first one?

3. (a) Can you work out the answer to these two?

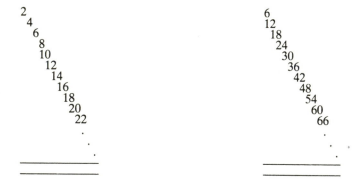

    (b) What is the connection between your two answers? Can you see two different connections between these answers and the answers you got in Question 2(a)?

4. Your answers to the first three questions were based, to some extent, on guesswork. It certainly *looks* as though the initial string of digits in each answer gets repeated over and over again. But it is not at all clear *why* this initial string of digits should go on repeating itself. So you cannot be sure that it will – especially since the further you go, the more difficult it gets to be sure that you have taken all possible 'carries' into account. Work out the next two numbers in each of these, then add them up.

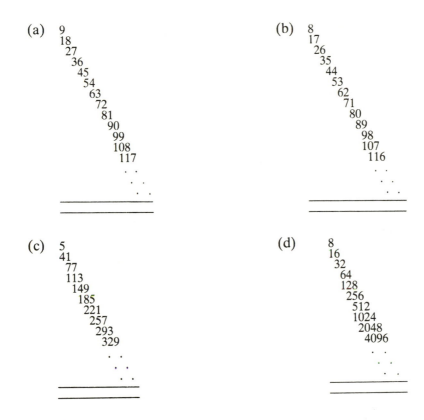

(a)
9
18
27
36
45
54
63
72
81
90
99
108
117
. .
. .
. .
_____

(b)
8
17
26
35
44
53
62
71
80
89
98
107
116
. .
. .
. .
_____

(c)
5
41
77
113
149
185
221
257
293
329
. .
. .
. .
_____

(d)
8
16
32
64
128
256
512
1024
2048
4096
. .
. .
. .
_____

5. (a) Where else have you met patterns of digits which repeat in the same way as in the answers to Questions 1–4?

(b) What are the advantages of interpreting the 'numbers' in Questions 1–4 as decimals whose decimal point has been accidentally left out?

6. (a) (No calculators please!) Work out the *exact* decimals for

$$\frac{1}{2} \quad \frac{1}{3} \quad \frac{1}{4} \quad \frac{1}{5} \quad \frac{1}{6} \quad \frac{1}{7} \quad \frac{1}{8} \quad \frac{1}{9} \quad \frac{1}{10} \quad \frac{1}{11} \quad \frac{1}{12}$$

(b) How can you tell without actually working it out whether a particular fraction, such as $\frac{1}{6}$, or $\frac{1}{27}$, or $\frac{1}{64}$, or $\frac{1}{320}$, has a decimal which goes on forever?

(c) Work out the exact decimals for

$$\frac{1}{7} \quad \frac{2}{7} \quad \frac{3}{7} \quad \frac{4}{7} \quad \frac{5}{7} \quad \frac{6}{7}$$

What do you notice? Does this suggest any connection between the answers to Questions 2(a) and 3(a)?

7. (a) Where have you seen the answer to this one before?

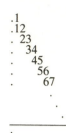

```
.1
.12
. 23
.  34
.   45
.    56
.     67
.
  .   .
  _____
  _____
```

(b) What are the answers to these two?

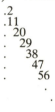

```
.2
.11
. 20
.  29
.   38
.    47
.     56

    .
   .
_____
_____
```

```
.1
.37
. 73
. 109
. 145
. 181
. 217

  . .
  . .
_____
_____
```

(c) Can you make up an endless sequence like '1,3,5,7, . . .' or '2,11,20,29, . . .' that will give the answer '0.222222 . . .' (the decimal for $\frac{2}{9}$)? What about '0.777777 . . .'?

8. (a) You know that 1.25 is the decimal for $1\frac{1}{4}$, and that 0.66666 . . . (forever) is the decimal for $\frac{2}{3}$. Which number has decimal 0.99999 . . . (forever)?

(b) Can you recognize the numbers whose decimals appeared in the answers to the endless sums of Questions 1, 2, 3, 4, 7?

(c) Work out the exact decimals for $\frac{1}{3}$, $\frac{1}{9}$, $\frac{1}{27}$, $\frac{1}{81}$.

9. The first few *Fibonacci numbers* are:

0, 1, 1, 2, 3, 5, 8, 13, 21, 34, 55, 89, 144, 233, ...

(a) Can you see how each Fibonacci number is calculated from the two numbers just before it? Use your rule to find the next two Fibonacci numbers after 233.

(b) The Fibonacci numbers go on forever. What is the answer to this endless sum? (You will have to be careful and patient!)

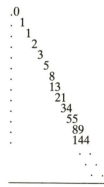

```
.0
. 1
   1
. 2
. 3
    5
. 8
  13
   21
    34
     55
      89
     144
  . .
   . .
  .
_____
_____
```

(c) Work out the exact decimal for $\frac{1}{89}$.(You will have to be careful and patient!).

Commentary

1. (a) **Make sure you have copied the sum down correctly,** shifting just *one* column to the right each time. And remember to add in columns as you always do. The first column adds up to 1; the second column adds up to 2; . . . ; the eighth column adds up to 8. *Oops!* We have forgotten the 'carry' from the ninth column; so the eighth column should be a __ . Carry on like this.

(b) You should be able to do this. But if you can't, don't worry – go straight on to Question 2.

2. (b) There is an obvious connection between
　　1, 3, 5, 7, 9, 11, 13, 15, . . . , and
　　3, 9, 15, 21, 27, 33, 39, 45, . . .
So how would you expect the two answers to be related?

3. (b) (i) There is an obvious connection between
　　2, 4, 6, 8, 10, 12, 14, 16, . . . , and
　　6, 12, 18, 24, 30, 36, 42, 48, . . .
So how would you expect the answers to be related?
(ii) There is an obvious connection between
　　2, 4, 6, 8, 10, 12, 14, 16, . . . , and
　　1, 3, 5, 7, 9, 11, 13, 15, . . .
There is a similar connection between
　　6, 12, 18, 24, 30, 36, 42, 48, . . . , and
　　3, 9, 15, 21, 27, 33, 39, 45, . . .
So how would you expect the answers in Question 3(a) to be related to those in Question 2(a).
(iii) Look hard and long at
　　1 3 5 8 0 2 4 6 9 1 3 5 8 0 2 4 6 9 1 . . . and
　　2 4 6 9 1 3 5 8 0 2 4 6 9 1 3 5 8 0 2 . . .
What do you notice?

5. (b) Whole numbers have a units column on the right hand end. The only numbers we know which do not have a fixed 'right hand end' are decimals. We could imagine the decimal point anywhere we like, but we shall always put it at the beginning. The second sum in Question 2(a) is then a lazy way of writing the sum shown here.

```
0.3
0.09
0.015
0.0021
0.00027
0.000033
─────────────
0.40740 . . . . . . . . . . . . . .
```

6. (a) Answers must be *exact.* If you stop writing before the exact decimal comes to an end, you must explain exactly how it goes on. The decimals for $\frac{1}{2}, \frac{1}{3}, \frac{1}{4}, \frac{1}{5}$ should present no problems. How can you work out the decimal for $\frac{1}{6}$ easily from one of these? The decimal for $\frac{1}{7}$ has to be worked out from scratch, but it is not hard. How can you work out the decimal for $\frac{1}{8}$ from one of the decimals you know already? And for $\frac{1}{9}$? (You only have to do one of the others 'from scratch'. Which is it?)

(b) (i) It is easy to jump to conclusions such as 'When n is even' or 'When n is a multiple of 2 or 5'. Always check such guesses. (What about $\frac{1}{6}$? Or $\frac{1}{15}$?)

(ii) What really matters is not whether n has 2 or 5 as a factor, but whether it has any *other* prime factors. Your final guess should be expressed in terms of 'the prime factorization of n', not just in terms of factors. It may help to think about the opposite question: What can you say about a number whose decimal does *not* go on forever? (What number has decimal 0.6? What number has decimal 0.62? What number has decimal 0.625? What is special about these numbers? Is $\frac{1}{64}$ special in the same way? How about $\frac{1}{320}$? Or $\frac{1}{360}$?

(iii) Why does the decimal for $\frac{1}{4}$ stop? 0.25 is the same as $\frac{25}{100}$. The decimal for $\frac{1}{4}$ stops after two places precisely because $\frac{1}{4}$ can be written as 'so many hundredths'. The decimal for $\frac{1}{80}$ stops after four places precisely because $\frac{1}{80}$ can be written as 'so many ten thousandths'. (How many?) For which values of n do we have $\frac{1}{n} = \frac{1}{10k}$ for some k?

(c) (i) When you wrote down the decimal for $\frac{1}{7}$ in part (a) you must have stopped writing before the end. How could you be absolutely sure that the first six digits in the answer go on repeating themselves forever? This should give you a clue how to write down the decimal for $\frac{2}{7}$ without doing any work. (The same applies to $\frac{3}{7}$, etc. . .)

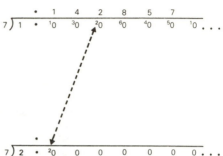

(ii) The repeating block of digits in the decimals for $\frac{1}{7}$ and $\frac{2}{7}$ are a bit like the repeating blocks of digits in the answers to the first part of Question 2(a) and the first part of Question 3(a). So perhaps the decimals in these answers correspond to fractions with the same _____ .

8. (a) It is natural to imagine that since 0.9 is just less than 1, and 0.99 is just less than 1, and 0.999 is just less than 1, so 0.9999 . . . . . (forever) must be just less than 1 too. **But decimals which go on forever are not like ordinary decimals which stop!** So how can we decide which number does have decimal 0.9999 . . . . . (forever)? Here are three different ways of trying to sort this out.

(i) 0.9 stops after one 9, and is less than 1 by 0.1 $= \frac{1}{10}$
0.99 stops after two 9's and is less than 1 by 0.01 $= \frac{1}{100}$
0.999 stops after three 9's, and is less than 1 by 0.001 $= \frac{1}{1000}$
But what if the 9's go on *forever*? How much smaller than 1 could the number be then?

(ii) Which fraction has decimal 0.111111 . . . . . (forever)? And what should we multiply by to get 0.999999 . . . . . (forever)? So which number has decimal 0.999999 . . . . . (forever)?

(iii) Let d = 0.999999 . . . . . (forever). Then 10 × d = 9.99999 . . . . . (forever). Subtracting we get

$$10 \times d = 9.99999 \ldots \ldots (\text{forever})$$
$$d = 0.99999 \ldots \ldots (\text{forever})$$
$$9 \times d =$$

so d = ____ .

(b) If you can't, part (c) should help!

(c) (i) There is no need to do any nasty calculations here. You know the decimal for $\frac{1}{3}$ is 0.3333 . . . . . (forever). To get the decimal for $\frac{1}{9}$ you just divide this by 3. And to get the decimal for $\frac{1}{27}$ you just divide again by __ . And to get the decimal for $\frac{1}{81}$ . . . . If you still have trouble identifying the numbers whose decimals appeared in the answers to Question 2, you might try multiplying your decimal for $\frac{1}{81}$ by 2, and by 11.

9.   (b)   You may have to find the first sixty or so digits in the answer before you can see what seems to be happening (so you will have to write down *more than* the first sixty *Fibonacci numbers*). When you think you know the answer, have a go at part (c).

(c)   You should be bursting to explain what turns up here. One way, which is not hard, is to use a variation of the standard trick used in the Commentary on Question 8(a)(iii). (The same standard trick should help in the Extension to Question 8.)

19

Counting: three

1. Start with a single equilateral triangle. At each stage add new equilateral triangles all round the outside.

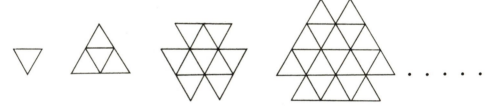

How many small triangles are there by the time you get to the 20th stage? The 200th stage?

2. (a) Start with a single square. At each stage add new squares all round the outside.

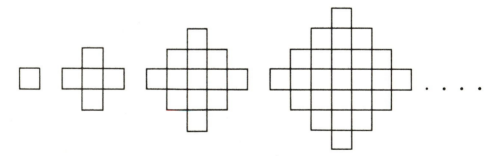

How many small squares are there by the time you get to the 20th stage? The 200th stage?

(b) Now do exactly the same starting with a single cube. At each stage add new cubes all round the outside. How many small cubes are there by the time you get to the 3rd stage? The 20th stage? The 200th stage?

3. (a) How many squares are hidden in this 4 by 4 grid? (The edges of the squares must lie along the grid lines.)

(b) How many squares would there be in a 40 by 40 grid?

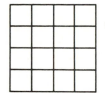

4. (a) How many squares are hidden in this 4 by 4 dot-lattice? (The corners of the squares must lie on the dots.)

(b) How many squares would there be on a 5 by 5 dot-lattice? How many squares would there be on a 50 by 50 dot-lattice?

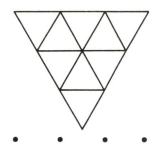

5. (a) How many equilateral triangles are there in this 3 by 3 triangular grid? (The edges of the triangles must lie along the grid lines.) How many would there be in a 4 by 4 triangular grid: How many in a 10 by 10 grid? Or a 25 by 25 grid?

(b) How many equilateral triangles are hidden in this 3 by 3 triangular dot-lattice? (The corners of the triangles must lie on the dots.) How many would there be in a 4 by 4 triangular dot-lattice? What about a 10 by 10 dot-lattice? Or a 25 by 25 dot-lattice?

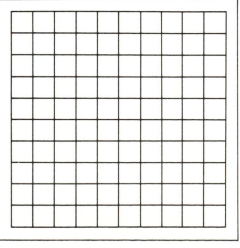

Investigation 1

(a) What is the least number of squares you have to remove from this 10 by 10 square grid so that no 2 by 1 rectangle can be fitted on what remains?

(b) What if you want to ensure that no 2 by 2 square can be fitted on what remains? What if no 3 by 3 square should fit on what remains?

(c) What if you want to ensure that no 3 by 2 rectangle can be fitted on what remains?

> **Investigation 2**
> How many different triangles can be made by choosing three rods
> from a collection of *n* straight rods of lengths 1, 2, 3, . . . , *n*?

Commentary

1. You could begin by just counting and making a table.

Stage: N	1	2	3	4	5	6	7	.	.	.
Number of triangles: T	1	4	10							

It should soon become clear how many triangles have to be added at each stage. This may help you to work out how many triangles there will be at the 20th stage. But to say how many there will be at the 200th stage, you need a formula giving the *number of triangles*, *T* in terms of the *stage*, *N*. (If you have difficulty producing such a formula, Section 17 may help.)

2. (a) Again you could begin by just counting and making a table like the one in the Commentary on Question 1. It should soon become clear how many squares have to be added each time, and you could use this to find out how many squares there will be at the 20th stage. But to say how many there will be at the 200th stage you really need a formula expressing the *number of squares* in terms of the *stage number*, *N*. Try to find such a formula on your own. (You can do this in the same way as in Question 1. A different method is indicated in the Commentary on part (b) below: don't read it until you get stuck on part (b).)

(b) Like all 3-dimensional problems this is much harder to picture than the corresponding problem for squares. But don't let that frighten you too much. If you try to draw up a table in exactly the same way as for the squares in part (a), you may well have trouble seeing how many cubes have to be added as early as the 3rd stage! So stand back and try to find another approach.

The art of solving many 3-dimensional problems lies in reducing them to 2-dimensional problems, for example, by looking at cross-sections. So instead of trying to count the number of cubes added all round the outside at each stage (a 3-dimensional problem!) and then adding up these 3-dimensional 'shells' or 'skins', try to find a good way of counting the total number of cubes at the *N*th stage by adding up the number of cubes in certain cross-sections (a 2-dimensional problem!). This should bring you back to the squares problem of part (a). (If you got stuck on part (a), you could try "shading" the squares at each stage as though they were on a chessboard. If this still doesn't seem to help, tilt your head slightly to one side!)

3. (a) Look for a systematic way of counting all the squares – from the one big 4 by 4 square down to the sixteen little 1 by 1 squares – which might reveal some significant pattern.

4. (a) You have already counted the 'obvious' squares – namely those with horizontal and vertical sides, like the ones in Question 3. But there are lots of others. As in Question 3, try to count them in a systematic way which might reveal some significant pattern in the number of squares of each type.

(b) The question asks about a 5 by 5 dot-lattice (and a 50 by 50 dot-lattice). If you think you can see what is going on, by all means tackle these right away (though you will have to count carefully and systematically, since there are over a hundred in the 5 by 5 dot-lattice). However you might find it helpful to think about a few easier cases first: if so, go back and recount the number of squares hidden in a 1 by 1 dot-lattice (1), in a 2 by 2 dot-lattice (6), and in a 3 by 3 dot-lattice (20), and use this experience to help you find a good way of counting. Then have a go at a 5 by 5 dot-lattice.

5. (a) If you count the small triangles only, you should not be surprised to find 3^2 in the 3 by 3 grid, 4^2 in the 4 by 4 grid, and so on. (Why?) But you are meant to count *all* the equilateral triangles, including those with sides of length 2, 3, etc. (And don't forget the upside down ones!) Counting the larger triangles may not suggest an obvious pattern. If it doesn't, look for a different way of counting. (One way which might help is to count the two types of triangles, apex up and apex down, separately. You should then see familiar looking numbers cropping up all over the place, but be careful with the 'apex up' ones! Another way is to consider odd and even grids separately.)

(b) You have already counted the 'obvious' triangles in part (a). Now you have to count the lopsided ones as well (provided that they are equilateral). Your experience with Question 4 should help you here. The problem may seem complicated at first, but don't let this put you off.

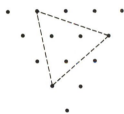

Investigation 1

(a) At first sight it may be a bit hard to see what this is getting at. There are two separate tasks: the first is to understand the problem sufficiently well to make an intelligent guess; the second is to explain why your guess is correct (or to improve it until it is correct). To understand a problem, it is always a good idea to look at smaller cases. How many squares do you have to remove from a 1 by 1 grid so that no 2 by 1 rectangle fits on what remains? How many squares do you have to remove from a 2 by 2 grid? (And *which* ones should you remove?) How many squares do you have to remove from a 3 by 3 grid? (And which ones should you remove?)

(b) At first sight this looks considerably harder. So start by experimenting on smaller boards. What is the least number you have to remove on a 1 by 1 or a 2 by 2 board? What about a 3 by 3 board? And so on. When you think you know what is going on, you will have to find some reason why your solution removes the smallest possible number of squares, no matter how large the board may be.

(c) This is harder. At first you will just have to find one way which works. (Removing all the 'white' squares would do, but it isn't very efficient.) Then try to *improve* on your first effort (that is, remove fewer squares while still ensuring that no 3 by 2 rectangle can be fitted on what remains). When you think you have found the most efficient solution, try to explain why it is impossible to do better.

Extension   How many 2 by 1 rectangles do you have to remove from an n by n square grid so that no 2 by 1 rectangle can be fitted on what remains?

20

Total information

1. (a) Suppose the six rectangular faces of a cuboid all have the same perimeter. What can you deduce about the cuboid? Explain why.

(b) Suppose the four triangular faces of a (not necessarily regular) tetrahedron all have the same perimeter. What can you deduce about the tetrahedron?

In Question 1 you were given information about certain 'totals', the perimeters of each face, and you had to try to extract information about the edge lengths themselves. All the problems in this section are like this, in that you are told something about certain totals, or averages, and have to deduce as much as you can about the original numbers.

2. (a) Place three *different* numbers in the circles so that the two numbers at the ends of each edge always add to a perfect square. What is the simplest possible solution?

(b) Place four *different* numbers in the circles so that the two numbers at the ends of each edge always add to a perfect square.

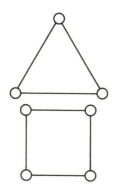

3. (a) Place four numbers in the circles so that the two numbers in the circles at the ends of each edge add to the number which is written on that edge.

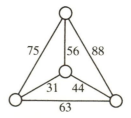

(b) Some of the ones below can't be done. Which ones? And why can't they be done?

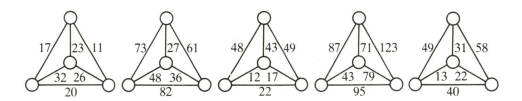

Can you find a way of telling *in advance* which ones can, and which cannot be done? And can you find a simple way of solving those which can be done?

4. (a) Place four numbers at the four vertices of this tetrahedron so that the three numbers at the vertices of each triangular face add up to the number which is written at the centre of that face.

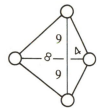

(b) Which of the puzzles below can, and which cannot be done?

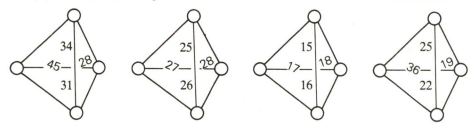

Can you find a simple way of solving those which can be done?

5. Any arrangement of the numbers 1–9 in the nine squares of a 3 by 3 array is called a 3 by 3 **magic square**, provided that the sum of the numbers in each of the three rows, in each of the three columns, and in each of the two diagonals are all equal.

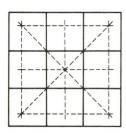

    (a) What must this sum be equal to?
    (b) Which number has to go in the central square?
    (c) What must opposite corners add up to?
    (d) Find all possible 3 by 3 magic squares.

6. Ten pupils sit an exam with their desks arranged in a circle. They all cheat so much that each pupil's final mark is the *average* of his/her two neighbours. What can you say about the ten final marks?

7. (a) The next year the same ten pupils sit another exam. This time their desks are arranged in a row. Unrepentant, they cheat just as much as before, with the result that the final mark of each pupil who has two neighbours is the average of these two neighbours' marks. What can you say about the ten final marks?

(b) Suppose infinitely many pupils sit an exam with their desks arranged in an (infinitely long) row. If they all cheat as before, and with the same effect, what can you say about their final marks?

8. (a) One hundred pupils sit an exam with their desks arranged in ten rows of ten. They all cheat so much that each pupil's final mark is just the average of his/her neighbours. (Pupil *A*, at a corner, has two neighbours. Pupil *B*, seated on an edge of the block, has three neighbours. Pupil *C* has four neighbours.) What can you say about their final marks?

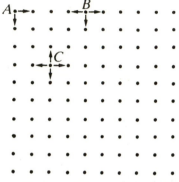

(b) What if there were infinitely many pupils with their desks arranged in an infinite square dot-lattice?

9. (a) I have three discs, each with one number on each side. Tossing the three discs together gives the following totals: 12, 13, 14, 15, 16, 17, 18, 19. When the total was 18, the numbers on the three visible faces were 5, 6, 7. Can you work out what numbers were written on the reverse sides? Is there just one solution, or are there several solutions?

(b) Is it possible, by starting with three suitably numbered discs, to obtain *any* eight consecutive numbers as totals in this way?

Investigation
(a) Choose any list of numbers. Replace each number which has two neighbours by the average of its two neighbouring numbers. Then do the same with your new list. Repeat this over and over again. What happens?
(b) Choose any square array of numbers. Replace each number by the average of its immediate neighbours. Then do the same with your new array. Repeat this over and over again. What happens?

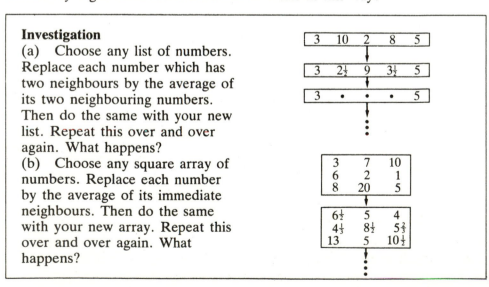

Commentary

1. (b) You *cannot* deduce that the tetrahedron is regular, but you can deduce quite a lot.

> **Extension** Suppose the eight triangular faces of a (not necessarily regular) octahedron all have the same perimeter.
> (a) What can you deduce about the octahedron?
> (b) What can you deduce if you are also told that each edge has the same length as its opposite edge?

2. (a) (i) Either the squares must be all even, or precisely two are odd. Can you see why?

(ii) Solutions with a 0 in one of the circles don't count! It is possible to find solutions with a 1 in one of the circles, though these are not really the 'simplest' solutions, since the numbers in the other two circles get rather large.

(b) The sum of the square on the top and bottom edges must equal the sum of the squares on the other two sides. Why? This should make your search a bit easier.

3. (b) What can you say about $a + d$? About $b + e$? About $c + f$? What sort of number must $d + e + f$ be? What can you say about $(a + b + c) - \frac{1}{2}(d + e + f)$? And how are $b + c$ and d related? (Which must be biggest?)

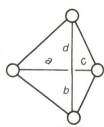

4. (b) What sort of number must $a + b + c + d$ be? What can you say about $a + b + c - 2d$?

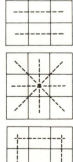

5. (a) What do you know about the sum of all three rows?

(b) What can you say about the sum of the four lines (horizontal, vertical, and two diagonals) which go through the centre square?

(c) What can you say about the sum of the two 'outside' rows and the two 'outside' columns?

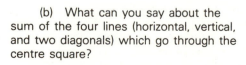

96

Extension 1 An arrangement of the numbers 1–16 in the sixteen squares of a 4 by 4 array is called a 4 by 4 **magic square** provided the four rows, the four columns, and the two diagonals have the same sum.
(a) What must this sum be?
(b) What must the four corners add up to?

(c) What must the four central squares add up to?

(d) What must the four squares in the middle of opposite edges add up to?

(e) Use this information to find a 4 by 4 magic square.

Extension 2 An arrangement of the numbers 1–27 in the twenty seven 'cells' of a 3 by 3 by 3 array is called a 3 by 3 by 3 **magic cube** provided the twenty seven rows and columns and the thirteen diagonals through the centre all have the same sum.
(a) Which number has to go in the central cell?
(b) Why must 1 and 27 go at opposite corners? Use the same idea to prove that it is impossible to construct a 3 by 3 by 3 magic cube in which *all* diagonals have the same sum – *face* diagonals as well as body diagonals.
(c) Use the ideas of Question 5 to find out as much as you can about the numbers which go in the other cells. Then construct your own 3 by 3 by 3 magic cube.

6. If you try attaching marks to the ten pupils you will decide fairly quickly what has to happen. You must then explain why it is bound to happen.
(i) Suppose pupil A scores M marks while pupil B on his left scores $M + d$ marks. What does pupil C on B's left score? And pupil D on C's left? If you carry on like this, what should pupil J on A's right score? So what must d be?
(ii) There is another observation which is often very useful in problems of this kind. Someone has to score the lowest mark! If there are several pupils with the lowest mark, choose any one of them. What can you immediately conclude about the marks scored by his/her two neighbours? What can you then conclude about all the other pupils' marks?
(iii) A third approach is to notice that the question has a certain *symmetry*. The desks are arranged in a circle. And the averaging condition applies equally to every pupil. So you should feel that there is no way any one pupil could possibly come out on top! As it stands this is scarcely a 'proof'; but you may agree that it does explain why the marks turn out as they do.

7. (a)   Each of the three approaches outlined in the Commentary on Question 6 can be used here – though with a different result this time. For example, the circular symmetry has been broken; but the line of thought outlined in part (iii) of the Commentary on Question 6 should still suggest who one might expect to come out on top (and who should come bottom!).

   (b)   With infinitely many pupils one cannot automatically assume that there is a lowest mark. (Why not? What if Pupil 1 got full marks, Pupil 2 got half marks, Pupil 3 got one third marks, and so on. Would there be a lowest mark?) But if infinitely many marks go up in equal steps – M, $M + d$, $M + 2d$, and so on – and the maximum possible mark is 100%, what can you say about the size of the equal steps?

8. (a)   You may feel that the question is a bit hard as it stands. One good way of beginning to get a feel for the problem is to start by looking at smaller arrays of desks – first 1 by 1, then 2 by 2, then 3 by 3, and so on, in the hope that this will give you an idea what happens for a 10 by 10 array of desks, and of how one might go about proving it.

   (i)   You can't say much about a 1 by 1 array.

   (ii)   In a 2 by 2 array, what can you say about the marks scored by pupils seated at opposite corners?

   (iii)   In a 3 by 3 array you should be able to write down lots of equations, and then combine them very nicely to see what has to happen.

   (iv)   In a 4 by 4 array this approach already begins to look too much like hard work. If you want to get up to a 10 by 10 array you are going to need another idea. Look back at the Commentary on Question 6 and see if any of the approaches described there look more promising.

   (b)   The subtle point we mentioned in the Commentary on Question 7(b) applies here too. As long as there are only finitely many pupils, at least one of them must score the lowest mark. If there are infinitely many pupils, then the same would still be true provided there were only finitely many different marks (as, for example, when each score has to be a whole number of marks out of 100). But if pupils are allowed to score *any* mark between 0 and 100 then there may not be anyone with the lowest mark, because there may not *be* a lowest mark. A different approach is needed.

9. (a)   There are several ways of tackling this. Most of them involve finding some suitable way of representing the data – in a picture, a table, or a diagram of some kind. One natural representation for a problem involving *three pairs* of numbers is to imagine the two numbers on each disc written on opposite faces of a cube. When you toss the three discs you get exactly one number from each of the three pairs; also, when tossing the three discs, there are exactly ____ possible outcomes. Thus, each outcome corresponds to the numbers on the three faces meeting at some ____ of the cube. It is therefore natural to picture the eight 'totals' 12, 13, 14, 15, 16, 17, 18, 19 attached to the eight ____ of the cube. You are told the numbers on the three faces which meet at the corner labelled 18. You have to choose the numbers on the other three faces so that the other seven corners produce the totals 12, 13, 14, 15, 16, 17, 19. There are several different solutions. Try to find them all.

   (b)   This is not as hard as it looks.

21

Eyeing the dots

You will need a supply of square dotty paper.

1. (a)   This 2 by 2 square has exactly one dot inside it. How many dots are there round the edge of the square?

(b)   Suppose we enlarge the original 2 by 2 square with scale factor 2, and place the enlarged square on the original array of dots (again with dots at the corners). How many dots will there be inside this enlarged square? And how many will there be round the edge of the enlarged square?

(c)   Suppose we enlarge the original 2 by 2 square with scale factor 3, and place the enlarged square on the original array of dots (again with dots at the corners). How many dots will there be inside this time? And how many will there be round the edge?

(d)   How many dots will there be inside, and round the edge, if we enlarge the original square with scale factor 10? Or 100?

2. (a)   What is the largest number of dots you can get inside a 2 by 2 square with dots at the corners? What if you are allowed to position the 2 by 2 square so that it no longer has dots at the corners?

(b)   What is the largest number of dots you can get inside a 5 by 5 square with dots at the corners? What if you are allowed to position the 5 by 5 square so that it no longer has dots at the corners?

3.   Find the exact areas of these shapes *without measuring*.

99

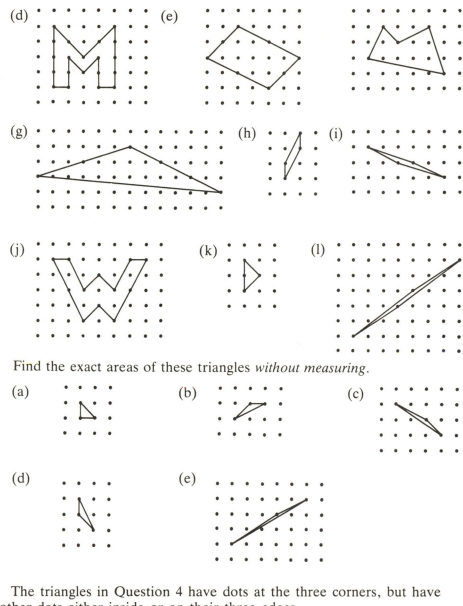

(d) (e)

(g) (h) (i)

(j) (k) (l)

4. Find the exact areas of these triangles *without measuring*.

(a) (b) (c)

(d) (e)

5. The triangles in Question 4 have dots at the three corners, but have no other dots either inside or on their three edges.

(a) It is sometimes difficult to tell 'by eye' from a sketch whether a particular dot is meant to be inside, bang on the edge, or outside the shape. Can you work out, just by looking at the three corners *A*, *B*, *C*, whether the triangle *ABC* will have any other dots either inside or on its edges?

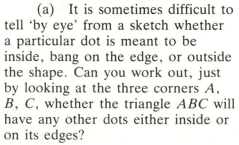
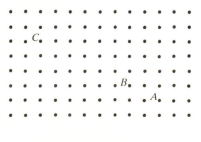

(b)   Suppose we choose coordinate axes so that the point A has coordinates $(0,0)$. What are the coordinates of the other two corners B, C?

(c)   Mark the five dots with coordinates $C = (-7,4)$, $F = (-5,3)$, $E = (-3,2)$, $D = (-1,1)$, $I = (1,0)$. What do you notice about the five points C, F, E, D, I? How is this line related to the line AB? What do you notice about CF and BA? What do you notice about FE and BA?

(d)   What can you deduce about the five triangles ABC, ABF, ABE, ABD, ABI?

(e)   What is the area of triangle ABD? So what is the area of triangle ABC?

6. (a)   Can you work out, just by looking at the three corners A, B, C, whether the triangle ABC will have any other dots either inside, or on its edges?

(b)   Suppose the point A, has coordinates $(0,0)$. What are the coordinates of B and C?

(c)   Find the coordinates of the point D such that CD and BA are parallel and equal in length. What do you know about the areas of triangles ABC and ABD?

(d)   Where have you seen a triangle like DBA before? What is its area? So what is the area of the original triangle?

7. (a)   If you had not already met a triangle like DBA in Question 6 it would not really have mattered. The way triangle ABD was obtained from triangle ABC indicates that we were thinking of the side AB as the base of the triangles ABC and ABD. If in the triangle ABD we switch our attention to the short side DB, and think of it as the base of the triangle DBA, then we can use exactly the same method as before to replace DBA by a simpler triangle. Keeping DB fixed, mark the points $A = (0,0)$, $E = (2,1)$, $F = (4,2)$, $G = (6,3)$. What do you notice about A, E, F, G? What does this tell you about the triangles DBA, DBE, DBF, DBG? And what is the area of the triangle DBG? So what was the area of the original triangle ABC?

(b)   Mark the triangle ABC where $A = (0,0)$, $B = (6,7)$, $C = (13,15)$. Find a slightly simpler triangle ABD with the same area as ABC. Then find a very simple triangle BDE with the same area as ABD. So what was the area of the triangle ABC?

(c)   Mark the dots $A = (0,0)$, $B = (7,9)$, $C = (13,17)$. Find a triangle ABD which is slightly simpler than, but which has the same area as, triangle ABC. Then find a very simple triangle BDE with the same area as ABD. Does the triangle BDE have any other dots either inside or on its edges? What does this suggest about the original triangle ABC? Go back and look carefully at the original triangle ABC to see whether it has any other dots either inside or on its edges.

8. Suppose ABC is any triangle with dots at each of its three corners, but with no other dots either inside or on its edges. Choose coordinate axes so that $A = (0,0)$, $B = (u,v)$, and $C = (x,y)$ (where u, v, x, y are whole numbers).

(a) Let $D = (x - u, y - v)$. What do you know about the lines BA and CD? What does this tell you about the triangles ABC and ABD?

(b) You know that triangle ABC has no other dots either inside or on its edges. Does it follow that triangle ABD has no other dots either inside or on its edges?

9. Questions 4–8 should have convinced you that a triangle ABC with dots at each corner but no other dots inside or on its edges always has area $\frac{1}{2}$. Can you use this fact to explain why a triangle which has dots at each corner and which has area $\frac{1}{2}$ *never* has any other dots either inside or on its edges?

10. (a) Divide each of the shapes in Question 3 into triangles which have dots at each corner but no other dots either inside or on their edges. Count the number of triangles you get for each shape, and compare this with the answer you found for the area in Question 3.

(b) Which pairs of shapes in Question 3 have equal areas?

11. Dividing shapes into little triangles is all good clean fun. But you should be wondering whether there is a better way of working out the areas of shapes like those in Question 3, which would tell you straight away when two different looking shapes, like those pairs in Question 10(b), have the same area, and which will help you work out the area of any given shape (like the shape **X** shown here) *in your head*.

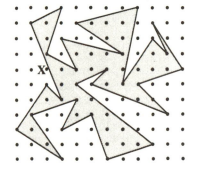

If there is a quick way of calculating areas, it will have to depend on easily available information. And the most obvious information about our shape is 'where the dots are': How many are there at the corners? How many are on the edges? How many inside the shape?

(a) Try to find your own quick way of calculating areas before looking up the Commentary on this part of the question.

(b) Use the quick method you found in part (a) to work out the areas of the shapes in Question 3 once more – this time *in your head*. Then work out the area of the shape **X** *in your head*.

12. In Questions 1 and 2 you saw that an N by N square in the 'obvious' position with dots at each corner has $(N - 1)^2$ dots inside it. What is the largest conceivable number of dots one could get inside an N by N square with dots at each corner? For which values of N is this largest conceivable number of inside dots actually possible?

13. Is it possible to choose three dots A, B, C in the square dot-lattice so that the triangle ABC is equilateral?

Investigation 1

In Questions 1 and 2 you saw that it is easy to position an N by N square on the square dot-lattice so that it contains N^2 dots inside it. For which whole numbers N is it possible to position an N by N square on the dot-lattice so that it contains *more than* N^2 dots inside it?

Investigation 2

(a)   Which of these triangles ABC have no other dots either inside or on the edges?

(i)   (ii)   (iii)

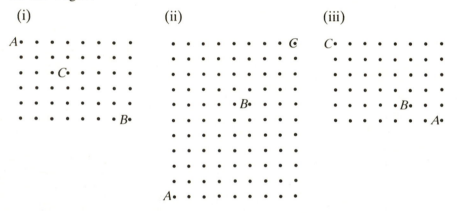

(b)   Can you find a simple rule which tells you straight away whether a triangle ABC with dots at each corner has any other dots either inside or on its edges?

Commentary

The supply of square dotty paper is important to allow you to experiment and 'get a feel' for each problem. This should help you see more clearly how to proceed.

1.   You are clearly meant to find an algebraic formula which expresses the number of dots inside, and round the edge, in terms of the scale factor.

2.   (a)   (i)   If the square does not have to have dots at the corners, then it is easy to get *four* dots inside. But can you see a way of possibly getting five dots inside? If you think you can, make sure you can prove that the dots are all *in*side, and not on the edge or outside the square.

(ii)   Is it possible to get six dots inside a 2 by 2 square? If you think you can, you should prove that they really are all inside. If you think it is impossible, try to explain *why* you think it is impossible.

(b)   (i)   A 5 by 5 square in the 'obvious' position with dots at the corners has 16 dots inside. But one can do a lot better than this by tilting the square – though you still have to find a position where the corners fit *exactly* on the dots.

(ii)   If the square does not have to have its corners on the dots, then it is easy to position it (with sides horizontal and vertical) to get 25 dots inside. Can you find a 'tilted' position for a 5 by 5 square which gives as many as 25 dots inside? Can you get more than 25 dots inside?

3.   All areas are to be measured using the area of one of the little squares as a unit. The shape in (a) has 15 square units. The other shapes are made up of *parts* of little squares. Your job is to find a way of combining these parts to obtain the exact area, *not* the approximate area.

5.   (a)   The only awkward dots are those labelled *D, E, F, G, H*. The line from *A* to *C* goes *up* 4 units and *back* 7 units (gradient $\frac{4}{-7} = -\frac{4}{7}$). The line from *A* to *B* goes up __ unit and back __ units. So is *B* above or below the line *AC*? And what about *D*? And *E, F, G, H*?

(d)   Same base *AB* and equal ____ , so equal ____ .

6.   (a)   See the Commentary on Question 5(a).
(c)   See the Commentary on Question 5(d).
(d)   Look back at the triangle *ABC* in Question 5(a).

7.   The important idea is to choose the shortest side as base and replace the given triangle by a simpler one with the same base and the same height, and hence the same area. Then choose the shortest side of the new triangle as a new base and repeat the whole procedure, going on like this until you get a *very* simple triangle whose area is:
(i)   easy to calculate, and
(ii)   equal to the area of the original triangle.

(b)   Choose *D* such that $\overrightarrow{AD} = \overrightarrow{AC} - \overrightarrow{AB} = (13,15) - (6,7) =$ ____ .
(c)   Choose *D* such that $\overrightarrow{AD} = \overrightarrow{AC} - \overrightarrow{AB} =$ ____ .

8.   (a)   Same base, equal heights, so same ____ .

(b)   *ABCD* is a parallelogram with dots at each corner. Triangle *ABC* forms exactly one half of the parallelogram, and *you know that there can be no extra dots in this half*. If there was a dot inside, or on the edge of, the new triangle *ABD* then it would have to be in the other half of the parallelogram – namely *CDA*. Suppose *E* were such an 'extra' dot.

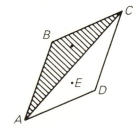

(i)   What sort of numbers occur as coordinates of dots? What does this tell you about the coordinates (*s,t*) of the dot *E*?

(ii)   Rotate the parallelogram *ABCD* through 180° about its centre. This maps the triangle *CDA* onto the triangle *ABC*, and the point *E* in *CDA* onto the point *E'* in the triangle *ABC*. What do you know about the coordinates (*m,n*) of the centre of the parallelogram? And what would this imply about the coordinates of the point *E'* in the triangle *ABC*?

9.   Suppose the triangle *did* contain other dots inside or on its edges. Then we could use these dots to cut up the triangle into (at least two) smaller triangles, each of which has dots at its corners but no other dots either inside or on its edges. What do you know about the area of each of these smaller triangles?

11.   (a)   Imagine a shape *X* with straight edges and with a dot at each corner. Let the area of the shape *X* be *A* (square units).

(i)   If you cut up the shape *X* into little triangles, each of which has a dot at each corner but no other dots either inside or on its edges, how many little triangles will there be?
Now work out the sum *S* of all the angles in all the little triangles *in two different ways*.

(ii)   First the easy way. There are ____ little triangles, and the angles of each little triangle add up to ____ , so
$$S = \_\_ \times \_\_$$

(iii)   Next the hard way.   Which dots occur as the corners of these little triangles? Precisely those which are inside, or on the edges of, the shape *X*! Let
   I = number of dots inside *X*,
   C = number of dots at corners of *X*,
   E = number of dots on edges, but not at corners, of *X*.

Each 'inside' dot will be completely surrounded by little triangles, and so will contribute ____ to the sum *S* of all the angles in all the little triangles. Hence the inside dots contribute *I* × __ in all. Each 'edge' dot is exactly 'half surrounded' by little triangles, and so will contribute ____ to the sum *S*. Hence the edge dots contribute *E* × __ altogether. And what about the angles at the 'corner' dots? You should know, or be able to work out, a formula for the sum of all the corner angles in a polygon with *C* corners.

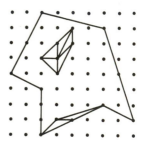

$$\therefore S = I \times \_\_\_ + E \times \_\_\_ + \_\_\_ \times 180 \text{ (from (iii))}$$
$$= 2A \times 180 \text{ (from (ii))}$$

This should give you a very easy formula for the area in terms of the number *I* of inside dots and the total number *E* + *C* of dots round the edge of the shape *X*.

12. The shape *X* here is an *N* by *N* square. What does this tell you about the maximum possible value of *I*?

Extension Can you find a way of predicting the largest number of dots one can actually get inside an *N* by *N* square with dots at each corner *just by thinking about the number N?*

13. An equilateral triangle with dots at each corner can be cut up into 'little triangles', each of area $\frac{1}{2}$. So its area would have to be a fraction. How does this help?

Investigation 1
Can you get two dots in a 1 by 1 square? Can you get three dots in a 2 by 2 square? Can you get ten dots in a 3 by 3 square? (Can you see how to get thirteen dots inside a 3 by 3 square?) Can you get seventeen dots in a 4 by 4 square? Can you get twenty six dots in a 5 by 5 square? (Can you see how to get thirty two dots inside a 5 by 5 square?)

Investigation 2
(i) Find a simple formula for the area of the parallelogram *ABCD* where $A = (0,0)$, $B = (a,b)$, $C = (c,d)$, $D = (a + c, b + d)$ are dots, (so *a, b, c, d* are whole numbers). Hence find a simple formula for the area of the triangle *ABC*.

(ii) Suppose the triangle *ABC* has some other dots on the edge *AB*. What would this tell you about the two coordinates *a* and *b*? And what would this imply about the number $ad - bc$? (Could $ad - bc = 1$?)

(iii) Suppose the parallelogram has dots inside it. What would this tell you about the number $ad - bc$?

(iv) It follows that if $ad - bc =$ ____ , then the triangle *ABC* cannot possibly have any other dots inside or on its edges. Is it also true that whenever the triangle *ABC* has no other dots either inside or on its edges, then $ad - bc$ has to equal ____ ? Explain!

22

Rectangles and squares

1. A block of chocolate consists of twenty-four small squares arranged in four rows of six. You wish to break it into its twenty-four small squares by making clean, straight, breaks – each break making two pieces out of one. What is the smallest number of breaks you need to make?

2. A 9 by 12 rectangular piece of paper is folded so that two opposite corners coincide. How long is the crease?

3. Here is a rectangular slice of fruit cake with icing on top and down the side. You have to share it equally with one other person. The icing is thicker down the side than on top. Can you divide the fruit cake and the icing exactly in half with a single straight cut? (If you think it can be done, explain where the cut must go. If you think it can't be done, explain why not.)

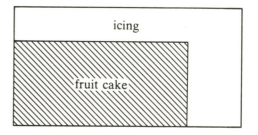

4. A square block of land fronts onto the street on all four sides. It has to be broken into five equal plots. The rateable value of each plot depends on two things only: its total area and its length of street frontage. Can you divide the block into five equal plots, all with the same rateable value? What is the simplest way of doing this?

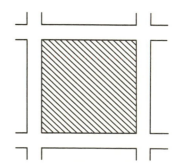

5. Show how to cut a square into exactly eleven squares. How many different ways can you find of doing this?

6. You have a square birthday cake and want to cut it up into acute-angled triangles. (In an acute-angled triangle, each angle has to be less than 90°.) Can it be done? If you decide that it cannot be done, explain why not. If you think it can be done, find the smallest possible number of pieces.

7. When we tile a rectangle with 2 by 1 rectangular tiles we sometimes get 'fault-lines'. These may run either vertically or horizontally.

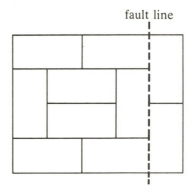

fault line

    (a) Show how to tile an 8 by 8 chessboard with 2 by 1 tiles without producing any fault-lines at all.

    (b) What is the smallest rectangle that can be tiled without producing any fault-lines at all?

    (c) What is the smallest square that can be tiled without producing any fault-lines at all?

Investigation 1

Imagine a square dot-lattice extending in all directions to cover the whole plane. Let the area of one of the basic squares be 1 unit.

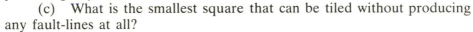

(a) Can you find a square with area exactly 2 units? (The corners of the square must lie on the dots.) Can you find a square with area 3? With area 4? 5? 6? 7? 8? 9? 10?

(b) For which numbers $N \leqslant 100$ is it possible to find a square of area N units? Can you find a general rule which will allow you to decide whether it is possible to find a square with area exactly 113, or 117, or 628, without even looking at the dot-lattice?

Investigation 2

It is not hard to convince yourself that a square cannot be cut into just two square pieces, or into just three square pieces. But it can obviously be cut into four square pieces (which happen to be all the same size). Is it possible to cut a square into exactly five square pieces? For which n is it possible to cut a square into exactly n square pieces?

Commentary

1. Each break increases the number of pieces by one!

2. There are several possible approaches – some more elegant than others. The first approach most people think of is to use Pythagoras' theorem several times over. But the fact that the answer turns out to be as simple as it does should suggest that there is a much simpler solution.

Extension *ABCD* is an *a* by *b* rectangle. *KL* and *MN* are two perpendicular 'diameters', (that is, lines which go through the centre of the rectangle). What can you say about *KL/MN*?

3. If the cut divides both the cake and the icing exactly in half, then it must also divide the whole rectangular piece exactly in half. Which straight lines divide a rectangle in half?

6. It would be a pity to give too much away here. After a while you may decide that it cannot be done: every new cut seems to produce another *obtuse*-angled triangle! If you think it cannot be done, you must try to explain *why not*. It is not enough just to feel convinced. (For example, one reason why many people decide that it cannot be done is because they only use cuts which go from one side right across to the other side, or from one side right across to a previous cut. This sort of cut is bound to produce one acute angle and one obtuse angle, or two right angles, at each end. But is this really the only possible kind of cut?) Stick at it!

Alternative Can you cut an obtuse-angled triangle into acute-angled triangles? If you decide that it cannot be done, explain why not. If you think it can be done, find the smallest possible number of pieces. (Does the size of the obtuse-angle matter?)

7. (b) (i) One way of approaching this is to think first about rectangles of width 2 (and any length). Is it possible to tile such a rectangle without fault lines?

(ii) Look next at rectangles of width 3 (and any length). Is it possible to tile such a rectangle without fault lines?

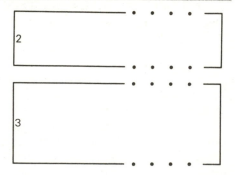

(iii)   What about rectangles of width 4?

(iv)   If a rectangle has width 5 and length less than or equal to 4, then by turning it on its side you would get one of the rectangles considered in (i), (ii), or (iii). Is it possible to tile a rectangle of width 5 and length 5 with 2 by 1 tiles?

(v)   What is the smallest rectangle that could conceivably be tiled with 2 by 1 tiles without fault lines? Can it actually be so tiled?

(c)   (i)   A square is a special kind of rectangle. Taking account of what you showed about rectangles in the Commentary on part (b) (i)–(iv), what is the smallest square that could conceivably be tiled with 2 by 1 tiles without producing fault lines?

(ii)   Can such a square actually be so tiled? Show how, or explain why not.

(iii)   So what is the smallest square that can be tiled with 2 by 1 tiles without producing fault lines?

Investigation 1

(b)   Choose any two dots P, Q in the dot lattice. Can you always find two other dots R, S so that $PQRS$ is a square? What is the area of such a square?

Investigation 2

(i)   Suppose the n square pieces were all the same size. What sort of number would n have to be?

(ii)   Suppose $n = 9$. Do the square pieces *have* to be all the same size?

(iii)   At some stage you should try to give a mathematical *proof* that it is impossible to cut a square into exactly two, or three, or five, or . . . square pieces.

23

Slices

The idea of this section is to get you to look for particular cross-sections of regular polyhedra. In the first few questions try to work with a *mental picture* of the polyhedron, and use your imagination to decide 'where to look' for the required cross-section. But at some stage you will need to make a sketch, or look at (and handle) a three-dimensional model of the polyhedron whose cross-sections you are trying to imagine.

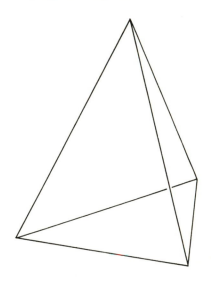

1. (a) How many different-sized equilateral triangular cross-sections can you get by slicing a regular tetrahedron? Describe precisely how to get such a cross-section. How can you be sure that these cross-sections really are *equilateral* triangles?

(b) Can you get a square cross-section by slicing a regular tetrahedron? How can you be sure that it is a square? How many different squares can you get? Can you get different-sized squares?

(c) Can you get a regular pentagon as a cross-section of a regular tetrahedron? Which other regular polygons can you get?

2. (a) Imagine a cube. What cross-section do you get when you slice it parallel to one of its six faces?

(b) There are infinitely many ways of slicing a cube parallel to one of its six faces. But they are not really very 'different', in that they all give exactly the same size and shape of cross-section in essentially the same way: namely a square exactly like one of the faces of the cube. Can you get a square cross-section by slicing the cube in a different way (that is, *not* parallel to one of the faces)? How many other ways are there of getting a square cross-section? Are any of these other square cross-sections bigger than or smaller than the square cross-sections you got in part (a)? Is it possible to get a different sized square cross-section by slicing a cube? (If you think it is possible, find one. If you think it is not possible, try to explain why not.)

(c)   You probably feel that you now know all there is to know about *square* cross-sections of a cube. You will have noticed already how to get lots of different-shaped *rectangular* cross-sections. Can you find a cross-section which is a parallelogram but not a rectangle? Can you get a rhombus which is not a square?

(d)   Can you get an equilateral triangle by slicing a cube? How can you be sure that it is *equilateral*?

(e)   Can you see how to get a regular hexagon by slicing a cube? How can you be sure that it is regular? How many regular hexagonal cross-sections are there?

(f)   Can you get a regular pentagon by slicing a cube?

3.   (a)   Can you get an equilateral triangle by slicing a regular octahedron? Can you get any triangular cross-sections at all?

(b)   Can you see how to get a square cross-section? How many different-sized square cross-sections can you get? How can you be sure that they are really *square*? What is the largest square cross-section you can get?

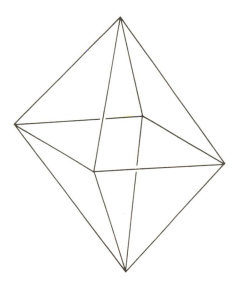

(c)   Can you see how to get a regular hexagon as a cross-section? How can you be sure it is regular? How many regular hexagonal cross-sections are there?

(d)   Which other regular polygons can you get by slicing a regular octahedron? Try to explain why you think you have found them all.

4.   Which regular polygons can you get as cross-sections of a regular icosahedron (twenty faces)? (Can you get a regular hexagon? What is the largest regular polygon you can get?)

5.   Which regular polygons can you get as cross-sections of a regular dodecahedron? (Can you get a square? What is the largest regular polygon you can get?)

6. On my desk I have a regular tetrahedron which fits exactly in a tetrahedron-shaped container on a stand. I want to raise the tetrahedron, turn it through 120°, and replace it in its container. How far must I raise the tetrahedron to do this?

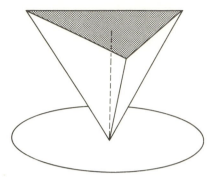

Investigation
(a)   Take a cube. Imagine all possible cross-sections of all possible shapes. What is the *largest* possible cross-section you can get?
(b)   Try to answer the same question for the other regular polyhedra.

Commentary

1. (a)   (i)   What happens if you slice parallel to the base?
(ii)   Suppose you rotate the tetrahedron about a vertical axis through the apex. What angle of rotation leaves the tetrahedron looking exactly the same as at the start? What does this tell you about the sides of the cross-section you got in part (i)?
(b)   (i)   How many edges are there in a regular tetrahedron?
(ii)   These edges come in three 'opposite' pairs. Imagine one opposite pair of edges drawn in red. How many other edges are there? What do you get if you slice the tetrahedron across these other four edges? Can you get a square? Is it really a square? (Remember, if you want to show that a quadrilateral is actually a *square*, it is *not* enough to show that its four sides have the same length. You have to show that all the angles are equal as well.)
(c)   You get a cross-section by slicing a solid with a plane. Where the slicing plane cuts a *face* of the tetrahedron we get a *side* of the cross-section. How many faces has the tetrahedron got? So what is the largest number of sides a cross-section could have?

2. (b)   (i)   How many edges are there in a cube?
(ii)   These edges come in six pairs of 'opposite' edges. Imagine the cross-section you get by slicing the cube *along* a pair of opposite edges. What shape cross-section do you get?
(iii)   If you imagine cross-sections parallel to this one, you get lots of other rectangles, all of the same height. Can you find one which is square?
(e)   The sides of the cross-section correspond to those faces of the solid which are cut by the slicing plane. So for a hexagonal cross-section of a cube the slicing plane has to cut all six faces. At first sight this may seem impossible but it is not.
(f)   It is certainly possible to get lots of pentagons, but none of them seem to be *regular*. It is not obvious how to prove that one cannot get a regular pentagon. Here is one approach.

(i)   For a pentagonal cross-section the slicing plane must miss one of the six faces altogether. Why?

(ii)   Call this face F. The cutting plane must cut at least one (in fact at least three) of the four parallel edges which have one end at a corner of F. Can you explain why?

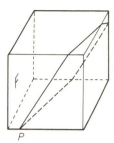

(iii)   One of the points where the slicing plane cuts these four edges is closest to F. Call this point P. Two sides of the pentagonal cross-section meet at P and both these sides must be directed 'away from' the face F. Why?

(iv)   What does this tell you about the angle at P? How big should the angle be at a corner of a regular pentagon?

3.   (a)   If you can imagine slicing 'right along' one of the faces (in which case the octahedron is not really cut into two separate bits!), then the cross-section would be an equilateral triangle. But can you get an equilateral triangle in any other way?

(d)   A cross-section cannot have more *sides* than the octahedron has *faces*. So the only remaining questions are whether one can get a regular pentagon, a regular heptagon (7 sides), or a regular octagon (8 sides) as a cross-section.

(i)   Does a regular octahedron have any octagonal cross-sections?

(ii)   Does a regular octahedron have any heptagonal cross-sections? (If you get stuck, look at part (iii) of the Commentary.)

(iii)   Show that for a pentagonal cross-section the slicing plane must cut through a corner of the octahedron. (Hint: imagine the faces of the octahedron coloured black and white so that adjacent faces have different colours. Use this colouring to show that for a cross-section with an odd number of sides, the slicing plane must cut through a corner of the octahedron.) What is the largest possible angle between the two sides of such a cross-section which meet at a corner of the octahedron?

4.   (i)   It is possible to get a cross-section with six equal sides, and another with nine equal sides. Can you see how to get such cross sections? Are they regular?

(ii)   There is one regular polygonal cross-section with more than nine sides. Can you find it?

5.   The dodecahedron seems to have the most regular polygons as cross-sections. It has all those that the icosahedron has, and two others besides.

6.   When you have found the answer, it should make you think again. (In particular, what has the problem got to do with 'Slices'?)

24

Prime time: two

1. In each of these lists there are exactly three numbers which are not prime numbers. Which are they?

(a)   11, 31, 41, 61, 71, 91, 101, 131, 151, 161, 171, 181, 191.

(b)   13, 23, 43, 53, 73, 83, 103, 113, 143, 153, 163, 173, 183, 193.

(c)   17, 37, 47, 67, 97, 107, 117, 127, 137, 147, 157, 167, 187, 197.

(d)   19, 29, 59, 79, 89, 109, 119, 139, 149, 169, 179, 189, 199.

2. Find two prime numbers which between them use each of the four digits 2, 4, 5, 7 exactly once.

3. Suppose you want to know whether the number 247 is a prime number.

(a)   Do you really have to try every number up to 246 to see whether it is a factor? Which numbers *do* you have to try?

Check your answer by reading the Commentary before going on.

(b)   Well? Is 247 a prime number? (See if you can do the calculations in your head.)

4. The method you used in Question 3(b) to decide whether 247 is a prime number is surprisingly effective. In this exercise you will produce *all prime numbers less than 120* simply by excluding all multiples of 2, 3, 5, and 7!

(a)   Write out the numbers up to 119 in twenty columns like this.

```
  1   2   3   4   5   6   7   8   9  10  11  12  13  14  15  16  17  18  19
 20  21  22  23  24  25  26  27  28  29  30  31  32  33  34  35  36  37  38  39
 40  41  42  43  44  45  46  47  48  49  50  51  52  53  54  55  56  57  58  59
 60  61  62  63  64  65  66  67  68  69  70  71  72  73  74  75  76  77  78  79
 80  81  82  83  84  85  86  87  88  89  90  91  92  93  94  95  96  97  98  99
100 101 102 103 104 105 106 107 108 109 110 111 112 113 114 115 116 117 118 119
```

(b)   There are good reasons for *not* calling '1' a prime number. (See the remarks in the Commentary.) So cross it out.

(c) The first uncrossed number in your table is the first prime number. Put a ring round it, and then cross out all multiples of that number. (You should be able to do this with ten strokes of the pen.)

(d) The first uncircled number which is left in your table must be the second prime number. Why? Put a ring round it, and then cross out all multiples of that number. (It is possible to do this with eight strokes of the pen. Can you see how?)

(e) The first uncircled number which is left in your table must be the third prime number. Why? Put a ring round it, and then cross out all multiples of that number. (This time four strokes of the pen should suffice.)

(f) The first uncircled number which is left in your table must be the fourth prime number. Why? Put a ring round it, and then cross out all multiples of that number. (This time three strokes of the pen should suffice.)

(g) There should now be exactly twenty-six uncircled numbers in your table. All you know about them is that they are all less than 120, and that none of them has 2, 3, 5, or 7 as a factor. Why does that guarantee that these twenty-six numbers are all prime numbers? Put a ring round each one.

5. Suppose you wanted to extend your list of the thirty prime numbers less than 120 to find all prime numbers less than 160. After adding two further rows to your table

120 121 122 123 124 125 126 127 128 129 130 131 132 133 134 135 136 137 138 139
140 141 142 143 144 145 146 147 148 149 150 151 152 153 154 155 156 157 158 159

you would obviously have to start by crossing out all the numbers in these two rows which are multiples of 2, 3, 5, or 7.

(a) Which other multiples would you have to cross out?

(b) How many of the numbers in these two rows are prime numbers?

6. (a) Go back to your table of prime numbers less than 120. Some of these prime numbers are equal to the sum of two squares (such numbers are sometimes called **sots**): for example, $2 = 1^2 + 1^2$, and $5 = 1^2 + 2^2$. Others cannot be written as the sum of two squares: for example, 3 and 7. Check each of the thirty prime numbers less than 120, one at a time, and put a 'box' round each one which can be written as the sum of two squares.

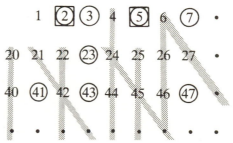

(b) Now take a long hard look at the prime numbers you have just 'boxed', and at the prime numbers you have left 'unboxed'. Suppose you allow for the fact that '2' is something of an exception – being the only even prime number (in many respects '2' is the 'oddest' prime number of all). Can you see a pattern beginning to emerge? (If not, it may be advisable to check your working in part (a) and part (b).) Where does this suggest you should expect to find prime numbers which are equal to the sum of two squares? And where would you expect to find prime numbers which are *not equal* to the sum of two squares?

(c)   Go through the prime numbers between 120 and 159 and mark all those you *expect* to be equal to the sum of two squares. Then check to see whether your guess was correct.

With luck you now have a fairly clear idea which prime numbers seem to be equal to the sum of two squares, and which do not. But how can you be sure that your guess is correct?

The first thing to notice is that you have actually made two quite distinct guesses. On the one hand, you probably expect that any prime number which is one more than a multiple of 4 must be equal to the sum of two squares. And on the other hand, you think that any prime number which is one less than a multiple of 4 cannot possibly be equal to the sum of two squares. It so happens that one of these guesses is easy to prove, while the other is not. (See the Commentary on Question 6(c).)

Investigation
(a)   Write the numbers 1–120 in a square spiral as shown here. Where are the prime numbers? Where are those prime numbers which are equal to the sum of two squares? Where are the square numbers?
(b)   What happens if you start with some number other than 1 at the centre? What if you start with 0? Or 3? Or 41?

```
37 36 35 34 33 32 31
38 17 16 15 14 13 30
39 18  5  4  3 12 29  .
40 19  6  1  2 11 28  .
41 20  7  8  9 10 27  .
42 21 22 23 24 25 26 51
43 44 45 46 47 48 49 50
```

Commentary

3.   (a)   You probably realize that there is no need to try any number which is bigger than *half* of 247 (since any number bigger than 123 goes into 247 less than twice). It is often difficult to convince people that one in fact only has to try *six different numbers*. Here is why.

If 247 is not a prime number, then we must be able to factorize it as the product of two smaller factors – say *a* and *b*:

$$247 = a \times b$$

(i)   These two factors a and b cannot both be bigger than *the square root* of 247. Why is that?

(ii)   Suppose a is the smaller of these two factors.
Then $a \leqslant \sqrt{(247)} = 15.6 \ldots\ldots$. So **if 247 is not a prime number, then it must have a factor a, which is greater than or equal to 2 and less than or equal to 15.**
This shows that you only have to try the fourteen numbers 2, 3, 4, 5, 6, ... up to 15. If none of them is a factor of 247, then 247 would have to be a prime number.

(iii)   But the problem of deciding whether 247 is a prime number is even easier than this. You certainly have to try each of the six *prime* numbers 2, 3, 5, 7, 11, 13 to see whether any of them is a factor of 247. But you don't have to worry about the other eight numbers! Why not? (Suppose one of the numbers 4, 8, 10, 14 was a factor. What else would have to be a factor? And suppose one of the numbers 6, 9, 12, 15 was a factor. What else would have to be a factor?)

(b)   You should be able to decide whether 2, 3, or 5 is a factor without even shutting your eyes! The other three primes 7, 11, 13 are not much harder as long as you realise that you don't need to know the *answer* to the division — you only need to know whether there is a remainder at the end. You should try to do this in your head.

4.   (b)   Prime numbers are as important in mathematics as atoms are in chemistry, or elementary particles are in nuclear physics. There are lots of different ways of splitting up 24 as a product of other numbers — for example,
$$24 = 24 \times 1 = 8 \times 3 = 6 \times 4 = 4 \times 3 \times 2 = 6 \times 2 \times 2 = 6 \times 4 \times 1$$
But there is only one way of writing it as a product of *prime* numbers, namely
$$24 = 3 \times 2 \times 2 \times 2$$
We can change the order in which we write the prime factors 3, 2, 2, 2, but we cannot alter the fact that there will always be one 3 and three 2's. If we were to call 1 a prime number, there would be lots of other ways of splitting up 24 as a product of prime numbers (such as $24 = 1 \times 3 \times 2 \times 1 \times 2 \times 2 \times 1$). For this and other reasons it is convenient *not* to call 1 a prime number.

(c)   The multiples of 2 form ten vertical lines in your table.

(d)   (i)   The first uncircled number left is 3. If it was not a prime number it would have to have a smaller factor. The fact that you have not yet crossed it out shows that it does not have any smaller factor (other than 1).

(ii)   The multiples of 3 lie in eight sloping lines (such as 24, 45, 66, 87, 108 and 21, 42, 63, 84, 105).

(e)   The first uncircled number left is 5. The fact that you have not yet crossed it out shows that it does not have any smaller prime factor. It must be a prime number.

(g)   If a number less than 120 is not a prime number, then we must be able to factorize it as the product $a \times b$ of two *smaller* numbers a and b.

(i)   Suppose a is the smaller of these two factors. Explain why a must be less than the square root of 120.

(ii)   Why does this guarantee that any number less than 120 which is not a prime number has to be a multiple of 2, or 3, or 5, or 7?

5.   (a)   If a number less than 160 is not a prime number, then it must have a prime factor less than $\sqrt{(160)}$. The only other numbers to cross out are multiples of __ .

6.   (b)   (i)   In which columns do the boxes occur? Are *all* the prime numbers in these columns boxed?

(ii)   Which columns contain prime numbers but no boxes?

(c)   Suppose $p = x^2 + y^2$ is a prime number. Then x and y cannot both be even (Why not?). And x and y are both odd only in one exceptional case. (When?) In all other cases we may suppose that x is odd and y is even. So what remainder will you get when you divide x^2 by 4? And what remainder will you get when you divide y^2 by 4? What does this tell you about the prime number p?

25

The ravages of time

1. A retired city gent, hopelessly lost on a country walk, spotted a yokel sitting on a fence. His initial sigh of relief was soon overtaken by his deepseated mistrust of all country folk. He would have to ask the way, but how could he be sure that the man would not misdirect him? He decided to test the water with a harmless opening question.

'What day of the week is this my good man?' he asked. The yokel saw straight through the gent's patronising ploy. At first he was tempted to ignore the question and remain silent. But after a while he replied: 'When the day after tomorrow is yesterday, today will be as far from Sunday as today was from Sunday when the day before yesterday was tomorrow.' Suitably stunned, the city gent retired to lick his wounds. What day was it?

2. Diophantus was a Greek mathematician who lived around the third century AD. Some of his books on arithmetic and algebra survived and influenced many later mathematicians – from Fibonacci (1170?–1240?) during the early Renaissance, right up to Fermat (1601–1665), who wrote his famous 'Last Theorem' in the margin of his own copy of Diophantus' works. But we know little more of Diophantus' life than is contained in the following epitaph:

> This tomb holds Diophantus. Ah, what a marvel! And the tomb tells scientifically the measure of his life. God vouchsafed that he should be a boy for the sixth part of his life; when a twelfth was added, his cheeks acquired a beard; He kindled for him the light of marriage after a seventh, and in the fifth year after his marriage He granted him a son. Alas! Late-begotten and miserable child, when he had reached the measure of half his father's life, the chill grave took him. After consoling his grief by the science of numbers for four years, he reached the end of his life.

How long did Diophantus live?

3. 'Tell me, old man, how old is your son?'
'As many weeks as my grandson is days.'
'And how old is your grandson?'
'As many months as I am in years.'
'But how old are you?'
'Our three ages together make a hundred years.'
Can you find their exact ages?

4. Now I am twice as old as you were when I was as old as you are now. When you are as old as I am now, together we will be sixty-three years old. How old are we now?

5. Keith is twice as old as Joseph was when Keith was half as old as Joseph will be when Joseph is three times as old as Keith was when Keith was three times as old as Joseph. Their combined ages make forty-eight. How old are they now?

6. I inherited a perfectly made, and perfectly symmetrical pocket-watch. The hour points are all marked in the same way, there are no numbers on the face. Of course, the only time the hour-hand and the minute-hand coincide exactly over one of the hour marks is at 12 o'clock. So if I wait long enough I will eventually discover which of the hour marks stands for 12 o'clock. But there is a way of telling the time straight away without waiting for the next 12 o'clock. Can you work out how?

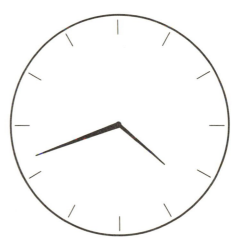

7. (a) How often each day do the minute-hand and hour-hand of a clock point in exactly the same direction?

(b) If at 6 o'clock we interchange the positions of the minute-hand and the hour-hand of a clock, we get an arrangement of the hands which never occurs in an accurate clock. How often do the two hands occupy positions which can be interchanged to give an arrangement which is correct for an accurate clock?

8. The hour-hand, minute-hand, and second-hand of a clock move smoothly and continuously. Can they ever be inclined at exactly 120° to each other?

Commentary

1. Try thinking in base 7.

3. It is tempting to start by letting the grandson be x days and the son x weeks old. Next you are told that the grandson is y months and the old man y years old. You are then faced with two (unavoidable) problems. First, how many days are there in a month? And second, an age of x weeks or y months does not usually take account of extra parts of weeks, or parts of months. But if you start at the other end, then it is natural to measure all ages in ____ .

4., 5. It would be a pity to spoil either of these. Your first solutions are unlikely to be very elegant. That doesn't matter. But don't be satisfied with just having got the answer. Go back and try to identify the crucial unknown which makes everything delightfully simple.

6. On the hour, the hour-hand points exactly to an hour mark, and the minute-hand points exactly to the 12 o'clock hour mark.

(i) At half past the hour, the hour-hand is exactly halfway between two hour marks. Where is the minute-hand?

(ii) How big is the angle $\alpha°$ between the hour-hand and the previous hour mark at twenty past the hour? At quarter to the hour? At twelve minutes past the hour?

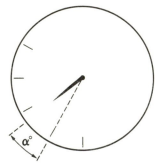

(iii) How many minutes past the hour is it when $\alpha = 12$? When $\alpha = 24$? When $\alpha = 9$? When $\alpha = 8$?

(iv) Can you see how to identify the 12 o'clock hour mark simply by measuring the angle $\alpha°$?

7. (b) It is worth looking for an 'Aha!' solution to this one. Imagine the hands permanently attached the wrong way round, with the hour-hand driven by the mechanism which should move the minute-hand, and vice versa. Your task is then to watch, and count the number of times the hands occupy positions which could occur in an actual clock. Every hour the hour-hand moves through each of the twelve 'hour spans' – 12 to 1, then 1 to 2, and so on up to 11 to 12 – while the minute-hand crawls through just 30°. If the hands were moving 'correctly', then as the hour-hand moves from one hour to the next, the minute-hand should start by pointing at the 12 o'clock mark and should then whizz right round so that it points at the 12 o'clock mark again just as the hour-hand reaches the next hour mark. In fact, the minute-hand creeps through a barely perceptible $2\frac{1}{2}°$ during this period! Compare the actual motion of the (barely moving) minute-hand with the (dizzily fast) motion which would be necessary if it were 'moving correctly' relative to the hour-hand. This should help you decide how often the interchanged hands occupy positions which could occur in an actual clock. You may find it helpful to represent the two motions graphically. (For an algebraic solution, see page 57 of the book *Algebra Can Be Fun* by Ya. I. Perelman, Mir Publishers.)

26

Chess pieces

1. (a) What is the largest number of 'peaceful' rooks one can place on a chessboard? (An arrangement of rooks is 'peaceful' if none of the rooks threatens a square which is occupied by another rook. The colour of the pieces is irrelevant.)

(b) How many different peaceful arrangements are there with this largest possible number of rooks?

2. (a) What is the largest number of peaceful bishops one can place on a chessboard?

(b) How many different peaceful arrangements are there with this largest possible number of bishops?

3. (a) What is the largest number of peaceful queens one can place on a chessboard?

∗(b) How many different peaceful arrangements are there with this largest possible number of queens?

4. What is the smallest number of rooks one has to use in order to threaten, or cover, every square on a chessboard? How many different ways are there of arranging this number of rooks so that they threaten, or cover, every square?

5. What is the smallest number of bishops needed to threaten, or cover, every square on a chessboard? How many different ways are there of arranging this number of bishops so that they threaten, or cover, every square?

6. What is the smallest number of queens needed to threaten, or cover, every square on a chessboard?

Investigation 1

You are allowed to start by placing a certain number of draughts, or checkers, on black squares below the central line. The only move allowed is 'jump-and-take' (in the diagram, *A* can jump over *B* and *B* must then be removed from the board). If you want eventually to get *one* draught up to 'level 1', you have to use at least two draughts.

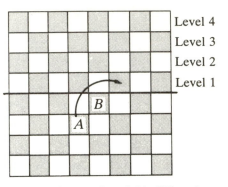

Level 4
Level 3
Level 2
Level 1

Suppose you want to get one draught up as far as 'level 2'. What is the least number of draughts you have to place on the board to start with? And how should you arrange them? What if you want to get one draught up as far as 'level 3'? 'Level 4'? 'Level 5'? . . .

Investigation 2

(a) What is the largest number of peaceful kings one can place on an *n* by *n* board?

(b) What is the largest number of peaceful knights one can place on an *n* by *n* board?

(c) What is the smallest number of kings needed to threaten, or cover, every square on an *n* by *n* board?

(d) What is the smallest number of knights needed to threaten, or cover, every square on an *n* by *n* board?

Commentary

1. (a) Each rook threatens one whole row and one whole column. After you have placed one rook, there are exactly 7^2 unthreatened squares. If you place a second rook on one of these unthreatened squares, how many squares will be threatened by neither of the first two rooks?

(b) (i) Can you explain why there has to be one rook in each of the eight horizontal rows? (Could there be two rooks in a single row? What if you had no rooks at all in, say, the bottom row?)

(ii) Suppose you wish to arrange eight rooks 'peacefully' on a chessboard. When you put the first rook in the first horizontal row, how many squares are there to put it in? And when you put the second rook in the second horizontal row, how many squares in that row are already threatened by the first rook? So how many squares are available for the second rook? So how many different ways are there of positioning the first two rooks in the first two rows? Carry on like this.

2. (a)   The diagonal moves of a bishop on a square chessboard are a bit more subtle than the horizontal and vertical moves of a rook. So you may not see through Question 2(a) as quickly as you saw through Question 1(a). But the two questions are really very similar!

(i)   A rook moves horizontally and vertically. So in Question 1(a) you could not put more than one rook in each horizontal row. This meant that the maximum number of peaceful rooks was at most eight. Question 2 is about bishops, which move diagonally. There are two kinds of sloping diagonals for bishops, those which run SE–NW and those which run NE–SW (just as there are two kinds of rows for rooks, horizontal and vertical). How many diagonals are there of each kind?

(ii)   Concentrate on one kind of diagonal. How many bishops can you place on each of these diagonals? (Careful! There are two of these diagonals which cannot *both* contain a bishop. Which are they?) What does all this tell you about the maximum number of peaceful bishops?

(iii)   Now you know that the number of peaceful bishops is at *most* __ , you must either find a way of placing exactly that many peaceful bishops on a chessboard, or else find some reason why it cannot be done.

(b)   In Question 1(b), provided you 'knew how to count' using the product rule and did not try to count the arrangements one at a time, you should not have had too much difficulty counting the $40320 = 8 \times 7 \times 6 \times 5 \times 4 \times 3 \times 2 \times 1$ different ways of aranging eight peaceful rooks on a chessboard. In contrast, it may seem impossible to count the number of different ways of arranging fourteen peaceful bishops on a chessboard. This is where mathematics comes into its own!

(i)   Whenever a problem seems too hard, go back and see what happens in simpler cases: in this instance, go back and see what happens on *smaller sized boards*.

Size of board	1 by 1 2 by 2 3 by 3 4 by 4 5 by 5 . . .
Largest number of peaceful bishops	1 2
Number of different arrangements	1

(Remember, there is no point filling in the bottom row of this table with wrong answers, or you will never see the light. So work carefully!)

(ii)   It would be very difficult to *count* (as opposed to *guessing*) the number of different arrangements on a large board if the bishops could go almost anywhere. Look back at the different arrangements you found on a 3 by 3, a 4 by 4, and a 5 by 5 board. Where do the bishops always seem to land up? (You should at least notice that there are certain parts of the board where the bishops *never* go.) Can you explain why this happens?

(iii)   At this stage you may not be able to explain why the bishops have to go where they do go. Don't worry. Just assume for the moment that in any arrangement of fourteen peaceful bishops on an ordinary chessboard, the bishops really do go where you think they have to. (You can come back later to try to explain why this has to happen.) Now see if you can use this 'fact' to help you to count (rather than guess) the number of different arrangements on an 8 by 8 board. (The actual answer is much simpler than the answer to Question 1(b).)

(iv)   Just in case you are still stuck, take a look at each of the eight squares in the top row in turn. Start with the top left square. There is just one other 'edge square' which lies on a diagonal with this. Where is it? You can place a bishop on just one of these two squares. So how many different ways are there of placing the

124

maximum number of peaceful bishops on this diagonal? Now look at the second square in the top row. There are two other edge squares which lie on diagonals with this square. Where are they? And there is one square in the bottom row which lies on diagonals with the same two edge squares. Where is it? These four edge squares (one in the top row, one in each side, and one in the bottom row) form the four corners of a tilted rectangle. How many peaceful bishops can one place on these four corners? Where do you have to put them if they are to be peaceful? So how many different ways are there of placing two peaceful bishops on two of these four squares? Now look at each of the remaining squares in the top row in turn.

    (v) And just in case you never discovered why the bishops have to go round the edge, you should now count the total number of threatened squares in two ways. *Concentrate first on the bishops.* How many squares are threatened by a bishop on an edge square? And what can you say about the number of squares threatened by a bishop which is not on an edge square? The next move is to add up the total number of squares threatened by all fourteen of the peaceful bishops. (Some squares will get counted twice, but don't worry.) What can you say about this total number of threatened squares if one or more bishops is on an inside square rather than an edge square? *Now turn your attention to the squares*, and ask yourself how many times each square can be threatened. Work out the maximum possible number of threatened squares and compare this answer with the previous one.

    3. (a) (i) A queen can do everything a rook can do. So the maximum number of peaceful queens cannot be greater than the maximum number of peaceful rooks.

    (ii) If you cannot see where to go from here, you may get some help by looking at what happens on smaller sized boards, but don't be misled by what happens on 2 by 2 and 3 by 3 boards! (Looking at simpler cases is a good habit. But *very* small cases sometimes fail to conform to the general pattern. For example, you may already have noticed that the largest number of peaceful bishops on a 1 by 1 board is larger than you would expect on the basis of the pattern which seems to hold for larger boards.)

    (b) A queen can do everything a rook can do *and* everything a bishop can do. So the number of arrangements will be less than the number of arrangements in Questions 1(b) and 2(b). (In fact it is much, much smaller.) As always, you should get some ideas by looking systematically at what happens on 4 by 4, 5 by 5, and 6 by 6 boards.

    4. After playing around for a while, you should feel that you have been here before.

    5., 6. After Question 4 you might expect these to be just like Question 2 and Question 3. But they aren't!

    (i) Start by looking at a 3 by 3 board.
    (ii) Then look at a 4 by 4 board.
    (iii) This should suggest that for an 8 by 8 board one can do considerably better than the fourteen bishops in Question 2 or the eight queens in Question 3.

Investigation 1
This is one of those instances where you should definitely notice a connection with an earlier section – though the connection is probably not quite as simple as you first imagine.

27

Round and round

1. (a) Given a list of numbers, such as '1,2,4', start at any point on the square grid, go 1 unit East, turn left through 90°, go 2 units North, turn left through 90°, go 4 units West, turn left through 90°, go 1 unit South, turn left through 90°, go 2 units East, and so on. What happens?

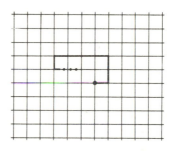

(b) What happens when you do the same sort of thing with each of these lists?

1,2,3,4;   1,2,3,4,5;   1,2,3,4,5,6;   1,2,3,3,2,1;
1,2,3,4,5,6,7,8,9.

(c) Make up some lists of your own. Do the same sort of thing with each list, and see what happens. *Try to predict what will happen before you start.* (What will happen to the list '1,2,3,4,5,6,7,8'? What about '1,2,3,4,5,6,7'? What about '2,4,6,8,7,5,3,1'?) *Then check to see if you were right.*

(d) When you think you have a rule which works, try to explain why it works.

2. (a) Given a list of numbers, such as '5,2', do exactly the same as in Question 1, but this time turn left through 120° between each step and the next. What happens?

(b) What happens if you do this for each of the six lists in Question 1(a) and 1(b)?

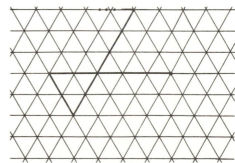

(c)   Make up some lists of your own. Do the same sort of thing with each list, and see what happens. Try to predict what will happen before you start. Then check to see if you were right.

(d)   When you think you have a rule which predicts correctly what will happen each time, try to explain why your rule works.

3.   (a)   Now do exactly the same as in Questions 1 and 2 but this time turn left through 60° between each step and the next. Try to predict what will happen with each list before you draw anything. (What will happen with the list '5,2'? What will happen with the list '6,5,3'? What will happen with '1,2,3,2,1,4'?) Check to see if you were right.

(b)   When you think you have a rule which predicts correctly what will happen each time, try to explain why your rule works.

Turns through 90°, or 120°, or 60° are very convenient because we can draw pictures so easily on a grid of squares or equilateral triangles. But there is no reason why we should not turn through any angle $\alpha°$ we choose. If you have access to suitable computer software (such as the language LOGO), you can draw pictures to help you explore what happens for other values of the turning angle $\alpha°$. If you do not have access to such aids, never mind! Rough sketches, imagination, and some mathematical thinking should do almost as well (though you will miss out on some pretty pictures).

Investigation
Can you find a rule which works for all values of the turning angle $\alpha°$? (As soon as you are told the value of the turning angle $\alpha°$, your rule should predict whether a given list of numbers will produce a path which joins up on itself, or a path which spirals further and further away from its starting point, or . . .)

Commentary

1 .   (a)   What happens after you have gone through the list four times? (Where are you? And in which direction will you move at the next step?)

(c)   You should find that most lists produce a path which joins up on itself. But some lists produce a path which spirals further and further away from its starting point. Can you see how to tell in advance what sort of path you will get *without actually drawing it first*? (Your first guess may not be quite right. It may predict well enough what happens with '1,2,3,4', or with '1,2,3,4,5,6,7,8'. But does it predict what actually happens with '1,2,4,3,6,3,3,2'?)

2.   **Make sure that you turn consistently through 120° rather than 60°.**

(a)   What happens after you have gone through the list three times? (Where are you? And in which direction will you move at the next step?)

(c)   You should find that most lists produce a path which joins up on itself, but that some lists produce a path which spirals further and further away from its starting point. You want a rule which accounts for what happens both with lists like '1,2,3,4,5,6', and with lists like '1,2,3,3,2,1' or '1,3,4,5,3,2'.

(d)   You should not have had too much difficulty explaining why your rule worked in Question 1. After all, the third step is in the opposite direction to the first, and the fourth step is in the opposite direction to the second. So one only has to make sure that at some stage the odd-numbered steps cancel out, and the even-numbered steps cancel each other out as well. It is tempting to use the same idea in Question 2 and to think in terms of the net horizontal and vertical displacements from the starting point. However, this time you never actually travel in the direction opposite to your initial direction, so the cancelling out is a bit more subtle. The approach can be made to work, though it is a bit messy. (And it gets worse for other values of the turning angle; see the Investigation.) You want to build on the experiences of Question 1 *without getting hooked on features which are peculiar to the case of a 90° turn*. Here are two things to think about which might help you to see Questions 1 and 2 in a different light.

(i)   What is the net effect of running through a list exactly once? (Where do you land up after one run-through? And in which direction are you facing?) What is the net effect of running through the list a second time? How is the net displacement on the second run-through related to the net dislacement on the first run-through the list?

(ii)   Suppose you add up all the displacements in each of the three separate directions. When will these three total displacements in the three separate directions cancel each other out exactly?

3.   With a 60° turn each time, you might be tempted to fall back on the 'opposite-directions-must-cancel-each-other-out' idea. This would explain why lists like '1,2,3,1,2,3' or '1,2,3,4,5,6,5,4,5,2,1,2' produce paths which join up on themselves. But your rule must also explain what happens with lists like '1,2,3,4,5,6'.

Investigation
It is worth sorting out first what sort of turning angles we should concentrate on.
(i)   Suppose $\alpha = \frac{m}{n}$ is a rational number (of degrees), where m and n are whole numbers with no common factor. After how many *turns* (through $\alpha°$ each time) will you be facing in exactly the same direction as at the start?
(ii)   How does your answer to part (i) tie in with what you found when $\alpha = 90$ (Question 1), when $\alpha = 120$ (Question 2), and when $\alpha = 60$ (Question 3)?
(iii)   Given your answer to part (i) it seems reasonable to hope that, provided $\alpha = \frac{m}{n}$ is a *rational* number (of degrees), every list will produce either a path which joins up on itself or a path which spirals further and further away from its starting point. All you have to do is to find out how to predict which, and to explain why your rule works.

Extension 1   Suppose α is an *irrational* number of degrees. Then you will never land up facing in the same direction as at the start. (Why not?) So the path you get may cross itself, but it will never join up on itself. What does happen when the turning angle is irrational? (For example, do you always spiral further and further away from your starting point? Or what?)

Those with access to LOGO or some other computer graphics facility will have a slight advantage in the above Extension, but they will have to use a little mathematical imagination since it is impossible to get an accurate irrational angle on a machine. The next Extension can be started by hand. But this time those with access to suitable computer graphics will have a considerable advantage. You should not expect to be able to explain much of what you find. But there is plenty of scope for you to investigate what happens, and to search for patterns.

Extension 2 We have, up to now, fixed the turning angle α and varied the length of each step. Now see what happens if you fix the step length (at, say, 1 unit or 5 units) and vary the turning angle each time in some agreed way. Consider, for example, the following sequences of instructions. (Each one is meant to go on forever.)

(a)   5 units East, turn 90° left, 5 units North, turn 180° left, 5 units South, turn 270° left, 5 units West, turn 360° left, 5 units West, turn 450° (= 90°) left, 5 units South, and so on.

(b)   5 units East, turn 60° left, 5 units forward, turn 120° left, 5 units forward, turn 180° left, and so on.

(c)   5 units East, turn 27° left, 5 units forward, turn 54° left, and so on.

(d)   5 units East, turn 29° left, 5 units forward, turn 58° left, and so on.

28

Prime time: three

'When he was about 15 years old, Gauss investigated the distribution of prime numbers. By counting those in each successive 1000 (1–1000, 1001–2000, 2001–3000, and so on) he eventually discovered a beautiful formula'

Do Section 24 first.
From Question 4 of Section 24 you already know that there are exactly twenty-five prime numbers between 1 and 100.

1. Extend the table in Question 4 of Section 24 up to 200. Then cross out multiples of 2, 3, 5, etc., (How far do you have to go?) and so find how many prime numbers there are between 100 and 200.

The aim of this section is to try to discover whether there is any rhyme or reason in the way prime numbers occur.

2. One approach to this question is to try to find some kind of formula for prime numbers. It is not easy to discover one formula which gives *all* prime numbers (and no other numbers) without missing any out. Still, have a go at these.

    (a) When $n = 0$, $n^2 + n + 17 = 17$, which is a prime number. When $n = 1$, $n^2 + n + 17 = 19$, which is also a prime number. What is the first value of n for which $n^2 + n + 17$ is not a prime number?

    (b) When $n = 0$, or $n = 1$, $n^2 - n + 41 = 41$, which is a prime number. What is the first value of n for which $n^2 - n + 41$ is not a prime number?

    (c) When $n = 0$, $n^2 + n + 41 = 41$, which is a prime number. What is the first value of n for which $n^2 + n + 41$ is *not* a prime number? Why should you have expected this?

For hundreds of years mathematicians thought it would be impossible to produce a single formula for all prime numbers. So it came as something of a surprise when various formulas were discovered in the 1970's. (These formulas are rather complicated and not very enlightening – see **Extension 2** to Question 2 in the Commentary.)

3. Another approach is to look for some unexpected bias in the way prime numbers occur.

(a) The only even prime number is 2. Every other prime number is odd, and so leaves remainder either 1 or 3 when we divide by 4: that is, either $p = 4n + 1$ or $p = 4n + 3$. Of the twenty-four odd prime numbers less than 100, how many have the form $p = 4n + 1$ and how many have the form $p = 4n + 3$? Of the twenty-one (odd) prime numbers between 100 and 200, how many have the form $p = 4n + 1$ and how many have the form $p = 4n + 3$?

(b) When you divide an odd number by 6, the remainder must be odd. Why? Which prime numbers leave remainder 3 when you divide by 6? So what do you know about the possible remainders when you divide a prime number $p \geqslant 5$ by 6? Of the twenty-three prime numbers between 5 and 100, how many have the form $p = 6n + 1$ and how many have the form $p = 6n + 5$? Of the twenty-one prime numbers between 100 and 200, how many have the form $p = 6n + 1$ and how many have the form $p = 6n + 5$?

(c) Of the twenty-four odd prime numbers between 1 and 100, how many have the form $p = 8n + 1$, how many have the form $p = 8n + 3$, how many have the form $p = 8n + 5$, and how many have the form $p = 8n + 7$? What are the corresponding numbers for the twenty-one primes between 100 and 200? And for primes between 200 and 300?

At first sight it is hard to see any rhyme or reason in the pattern of prime numbers. However, one can see some general trends.

4. There is a table of prime numbers up to 5000 at the back of the book.

(a) Use this table to count the number of prime numbers in each successive hundred, from 1–100 all the way up to 4901–5000. Make a table of your own.

Between:	1–100	101–200	201–300	301–400	. . . . .	4901–5000
Number of primes	25	21				

What general trends do you notice?

(b) Then make a list of the number of prime numbers in each successive 200, from 1–200 all the way up to 4801–5000.

Between:	1–200	201–400	401–600	601–800	. . . . .	4801–5000
Number of primes	46					

What general trends do you notice?

You should no longer expect to find a simple rule which describes exactly how prime numbers occur. You may even be beginning to wonder whether there is any rhyme or reason to prime numbers and the way they occur! But despite the persistent 'hiccups' which spoil the general trends you noticed in Question 4, there is in fact a surprising overall pattern to the way prime numbers occur. This is the 'beautiful formula' which Gauss (and others) discovered in the 1780's and 1790's.

5. To give you a chance of discovering this relationship for yourself, it is a good idea to count how many prime numbers occur up to certain very specific points: these are the powers of one particular number e, where $e = 2.7182818 \ldots \ldots$ (There is in fact nothing special about these numbers other than the fact that they make it easier for you to see what is going on!) Use the table of prime numbers up to 5000 to complete the following table. Then try to make some sense of what you find.

n	e^n	Next whole number N	Number of primes $\leqslant N$
1	2.718 . . .	3	2
2	7.389 . . .		
3	20.08 . . .		
4	54.59 . . .		
5	148.41 . . .		
6	403.42 . . .		
7	1096.63 . . .		
8	2980.95 . . .		
9	8103.08 . . .		1019

Commentary

1. Remember that you only have to cross out multiples of prime numbers up to $\sqrt{200}$: that is, multiples of 2, 3, 5, __, __, and __ . You should find just four fewer prime numbers than you did between 1 and 100.

2. (a) Look carefully at the expression $n^2 + n + 17$. Can you see a value of n which cannot possibly give a prime number?
　　(b) Look carefully at the expression $n^2 - n + 41$. Can you see a value of n which cannot possibly give a prime number?
　　(c) The two expressions $n^2 - n + 41$ and $n^2 + n + 41$ are related in a very simple way. Can you see how? (Why is the value of $n^2 - n + 41$ when $n = 11$ exactly the same as the value of $n^2 + n + 41$ when $n = 10$?)

3.   (a)   You should not expect the two numbers to be *exactly* equal. It is trends that interest us here. What do you think happens in the long run? Are there more primes of the form $p = 4n + 1$ or more of the form $p = 4n + 3$? Or are there roughly the same number of each kind?

4.   (a)   The first two entries '25' and '21' are considerably bigger than all subsequent entries. Why do you think this is? The number of prime numbers in each successive hundred tends to decrease – though not in any simple-minded way. (For example, there are just as many prime numbers between 4201 and 4300 as there are between 201 and 300!)

(b)   The general trend, 46, 32, 30, 30, 29, 28, 27, 28, 27, 24, 27, 25, 25, 26, 26, 23, 21, 26, 27, 23, 26, 25, . . . , should be clearer here (though there is still no simple-minded pattern).

Extension What happens if you count the number of prime numbers between 1–200, then 101–300, then 201–400, and so on? Do you get a steadier decrease? Or do 'hiccups' still occur? What if you count the number of prime numbers between 1–400, then 101–500, then 201–600, and so on?

5. (i) When looking for patterns like this, you will probably notice first that each number in the right hand column is somehow related to the one before it. The initial doubling is very striking, but it doesn't last. After a while the number of prime numbers starts going up by a factor slightly bigger than two. What do you think might be happening to the ratio of successive entries in the right hand column?

(ii) Don't be content merely to find this 'vertical' trend relating each entry in the right hand column to the next entry below it. Try to see some 'horizontal' connections relating the number of primes $\leq N$ to the number N, or e^n, or n. Use what you find to predict how many prime numbers you would expect there to be $\leq e^{10} \approx 22027$. (The actual number is given at the end of this Commentary.)

Question 5 is an attempt to say just how many (or rather, what proportion of) whole numbers $\leq N$ are prime numbers. A mathematical proof that the patterns you noticed in Question 5 really do continue for all values of $n > 9$ is not easy. It took mathematicians over a hundred years to discover 'the reason why'!

Behind Question 5 is the fact that, though prime numbers become less common as the size of the numbers increases, they never come to an end: the sequence of prime numbers is endless, or infinite. The proof of this basic fact has been known for over two thousand years. It is based on a very simple idea.

● 2 is the first prime – call it p_1. Now p_1 is not a factor of $p_1 + 1$. (If we divide by p_1 we get remainder 1.)

● Therefore each prime factor of $p_1 + 1$ is different from p_1. Let p_2 be the smallest of these prime factors. (In fact, $p_1 + 1 = 3$ only has one prime factor, so $p_2 = 3$. But this is not important.) Then neither p_1 nor p_2 is a factor of $p_1 \times p_2 + 1$. (If we divide by p_1, or by p_2, we get remainder 1.)

● Therefore each prime factor of $p_1 \times p_2 + 1$ is different from both p_1 and p_2. Let p_3 be the smallest of these prime factors. (In fact, $p_1 \times p_2 + 1 = 7$ only has one prime factor, so $p_3 = 7$.) None of p_1, p_2, p_3 is a factor of $p_1 \times p_2 \times p_3 + 1$. (Why?)

● Therefore each prime factor of $p_1 \times p_2 \times p_3 + 1$ is different from p_1, p_2, and p_3. Let p_4 be the smallest of these prime factors. (In fact, $p_1 \times p_2 \times p_3 + 1 = 43$ only has one prime factor, so $p_4 = 43$). None of p_1, p_2, p_3, p_4 is a factor of $p_1 \times p_2 \times p_3 \times p_4 + 1$. (Why?)

● Therefore each prime factor of $p_1 \times p_2 \times p_3 \times p_4 + 1$ is different from p_1, p_2, p_3, and p_4. Let p_5 be the smallest of these prime factors. (In fact, $p_1 \times p_2 \times p_3 \times p_4 + 1 = 1765$, so $p_5 = 5$.)

We can continue like this for ever. Admittedly, the procedure does not produce the prime numbers in their 'natural' order: 2, 3, 5, 7, 11, 13, etc. But it does produce an endless sequence of different primes, and so proves that there are infinitely many prime numbers.

(The number of prime numbers ≤ 22027 is 2467.)

29

A fitting end

This set of problems is mainly concerned with fitting regular polygons together edge-to-edge, leaving no gaps. In the resulting pattern the arrangement of polygons round a common corner is called a *vertex figure* of the pattern. For several of the problems I would encourage you to use plastic tiles which clip together (such as those produced with the trademark 'POLYDRON') as an aid to thought.

1. A tiling (or tessellation) of the plane is a *regular tiling* if the tiles used are all regular polygons of the same kind, and fit together edge-to-edge, leaving no gaps. The most familiar regular tiling is the arrangement of square tiles one finds on many bathroom walls. The vertex figures for this tiling are all congruent: each vertex figure consists of four squares meeting round a single common point.

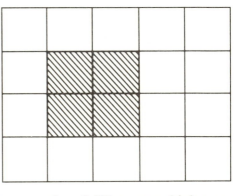

    (a) How many other regular tilings are there? When you think you have found them all, explain why there cannot possibly be any others.

    (b) In any regular tiling the vertex figures are all automatically congruent. Can you explain why?

2. A polyhedron is a *regular polyhedron* if it has finitely many faces which are all regular polygons of the same kind, fitting together edge-to-edge and leaving no gaps in such a way that the vertex figures are all congruent. The most familiar regular polyhedron is the cube.

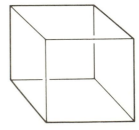

Each vertex figure of a cube consists of three squares meeting at a point. How many other regular polyhedra are there? When you think you know, explain why there cannot possibly be any others.

3.   A tiling of the plane is a *semiregular tiling* if the tiles are all regular polygons, though not necessarily all of the same kind, which fit together edge-to-edge leaving no gaps, in such a way that the vertex figures are all congruent. You can think of the three regular tilings you found in Question 1 as very special semiregular tilings in which the tiles all happen to be of the same kind.

The simplest semiregular tiling with tiles of more than one kind is perhaps that formed by putting together alternate strips of squares and equilateral triangles. Each vertex figure of this tiling consists of three triangles and two squares,

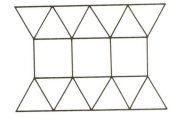

with the two squares always side by side and the three triangles always together. How many other semiregular tilings of the plane are there?

4.   A polyhedron is a *semiregular polyhedron* if its (finitely many) faces are all regular polygons, though not necessarily all of the same kind, which fit together edge-to-edge, leaving no gaps, in such a way that the vertex figures are all congruent. You can think of the five regular polyhedra you found in Question 2 as very special semiregular polyhedra in which the faces all happen to be of the same kind. How many semiregular polyhedra are there?

5.   (a)   A *regular tiling* of space is a tiling with regular polyhedra all of the same kind, which fit together face-to-face, leaving no gaps, and which fill up the whole of 3-dimensional space. How many different regular tilings of space are there?

   (b)   A *semiregular tiling* of space is a tiling with regular polyhedra, not necessarily all of the same kind, which fit together face-to-face, leaving no gaps, and which fill up the whole of space in such a way that the vertex figures are all congruent. How many different semiregular tilings of space are there?

Investigation
A polyhedron whose faces are all equilateral triangles is called a *deltahedron*. The faces of a deltahedron are all congruent, but the vertex figures need not be congruent.
(a)   How many deltahedra are there without 'dents' or indentations?
(b)   How many deltahedra can you find in which each vertex figure consists of at most six equilateral triangles?

Commentary

In several of the questions in this section you will need to know the size of each corner angle in regular polygons with various numbers of sides. It is worth making a list for regular polygons with up to, say, forty-two sides.

Number of sides	3	4	5	6	7	8	9	10	11	12	13	14	...	42
Size of corner angle	60	90												

1. (a) The sum of the angles which fit together round a point must add up to 360°. This restricts the kind of regular polygons which can be used in a *regular* tiling. It also determines how many of them fit together round each point.

(b) There must be at least three tiles meeting at each point. (Why couldn't there be just two tiles?) So the corner angle of each tile is at most ____ . It must also go into 360° a whole number of times.

2. (i) One way in which it is often 'proved' that there are only five regular polyhedra begins by observing that, whereas the angles round a point in a regular *tiling* must add up to *exactly* 360°, the angles round a corner in a regular *polyhedron* add up to *less than* 360°. Show that, if this assertion is correct, then there are indeed only five possible regular polyhedra.

(ii) The assertion on which the 'proof' in part (i) was based can be justified only if the polyhedron does not have any dents. It is perfectly possible to fit seven or eight (or more) equilateral triangles together round a point without gaps or overlapping provided one is allowed to produce a 'corrugated' surface. Similarly, it is perfectly possible to fit five or six (or more) squares together round each point. At least one of these 'corrugated' arrangements of polygons can be continued in such a way that its vertex figures are all congruent. Why does this not add to the list of regular polyhedra? One reason is that the arrangement of regular polygons does not close up to give a finite polyhedron, but stretches on forever and so consists of infinitely many polygonal faces.

To prove that the only *finite* regular polyhedra are the five you found in part (i) you need a proof which depends in some way on the fact that a polyhedron has only *finitely many faces*. In other words, you must count something. The best known formula for polyhedra which depends on counting is *Euler's* formula
$$V + F = E + 2,$$
where V is the number of vertices, F is the number of faces, and E is the number of edges of the polyhedron. (For the cube we have $V = 8$, $F = 6$, and $E = 12$.) The numbers V, F, E do not describe the shape of the polyhedron completely. (Can you find a polyhedron which has the same values of V, F, E as the cube, but which does not have six quadrilateral faces?) But no matter how bent or dented the polyhedron may be, the numbers V, F, E always satisfy *Euler's* formula provided that *the polyhedron is like a sphere* in the sense that if we were to remove all the dents by 'pumping it up', then it would turn out like a football, rather than like a bicycle tyre, or some other shape with one or more 'holes' through the middle.

(iii) Suppose a regular polyhedron is made of regular polygons having p sides, with q of these polygons meeting at each vertex. Explain why $F \times p = V \times q = 2E$.

(iv) Use this to substitute for F and V in *Euler's* formula. Hence show that

$$E = \frac{2pq}{4 - (p - 2)(q - 2)}$$

(v) The number of edges E must be positive, so $(p - 2)(q - 2)$ must be less than 4. You also know that p and q must both be at least 3. (Why?) So what are the possibilities for the pair p, q?

3. Perhaps the simplest way of tackling this is to use the same ideas as in Question 1, namely that the sum of the angles meeting at a point must be exactly 360°.
(i) Make a list of all the possible combinations of regular polygons which can fit together exactly round a point. There are twenty-one different combinations. (Different ways of arranging the same set of polygons count as different.)
(ii) Now try to see which of these vertex figures *cannot* be continued to give a semiregular tiling of the whole plane. (Luckily one simple idea suffices to exclude all those vertex figures which do not give a semiregular tiling. Notice that if the vertex figure consists of exactly three regular polygons, all of different kinds, they must all have an even number of sides. Can you see why? This means that only one of the six vertex figures with three different polygons which you found in part (i) could actually occur in a semiregular tiling. A slight variation of this idea shows that whenever the vertex figure consists of three or four polygons, one of which has an odd number of sides, the two polygons on either side of the odd polygon in each vertex figure must have the same *even* number of sides. This rules out five more of the vertex figures you found in part (i).) You should now check that each of the remaining eleven vertex figures gives rise to a semiregular tiling of the plane. (One gives rise to two different tilings, though they are mirror images of each other.)

4. There are two possible approaches here.
(i) The simple-minded approach is to restrict to polyhedra without dents. The sum of the angles at each vertex will then be less than 360°. You can then make a list of all the possible combinations of three or more regular polygons for which the sum of the angles is less than 360°. Then you can try to see which of these combinations could actually give rise to a semiregular polyhedron. (You will need to think a little before making your list of 'possible vertex figures'. Otherwise your list will be endless(!), since any regular polygon whatsoever can be combined with two equilateral triangles, or with an equilateral triangle and a square, or . . . to produce a vertex figure for which the sum of the angles is less than 360°.)
(ii) The second, more efficient, approach is to use *Euler*'s formula. Let the number of polygons in each vertex figure be f. Let F_i denote the number of faces of the polyhedron which have exactly i sides, and let f_i denote the number of faces in each vertex figure which have exactly i sides. (In a *regular* polygon the numbers F_i and f_i are equal to zero for all except one value of i.) Clearly

$f = f_3 + f_4 + f_5 + f_6 + \ldots$ and
$F = F_3 + F_4 + F_5 + F_6 + \ldots$

(1) Show that $3 \times F_3 + 4 \times F_4 + 5 \times F_5 + 6 \times F_6 + \ldots = 2E$
(2) Show that for each i, $V \times f_i = F_i \times i$ Deduce that $V \times f = 2E$
(3) By substituting in *Euler*'s formula, show that for each value of x we have

$$x(F_3 + F_4 + F_5 + F_6 + \ldots) = x \times F = x \times V\left(\frac{f}{2} - 1\right) + 2x.$$

(4) Use (3) to prove that for any semiregular polyhedron, and for each value of x, we have

$$(x - 3)F_3 + (x - 4)F_4 + (x - 5)F_5 + (x - 6)F_6 + \ldots = 2x + V\left(\frac{xf}{2} - x - f\right).$$

(5) Substitute $x = 3$ in (4) and hence show that in any semiregular polyhedron the number f of polygons at each vertex is 3, 4, or 5.

(6) Use (2) and divide both sides of (4) by V to show that

$$\frac{(x-3)}{3}\,f_3 + \frac{(x-4)}{4}\,f_4 + \frac{(x-5)}{5}\,f_5 + \frac{(x-6)}{6}\,f_6 + \ldots\ldots = \frac{2x}{V} + (\frac{xf}{2} - x - f).$$

(7) If $f = 5$, choose x so that $(\frac{xf}{2} - x - f) = 0$. Then solve the resulting equations

$$\tfrac{1}{9}f_3 - \tfrac{1}{6}f_4 - \tfrac{1}{3}f_5 - \tfrac{4}{9}f_6 - \ldots\ldots = \frac{20}{3V}$$

$$f_3 + f_4 + f_5 + f_6 + \ldots\ldots = 5.$$

(The first equation says that f_3 must take the lion's share of $f = 5$. There are two possible values for f_3 and three possible semiregular polyhedra – two of which have different mirror image versions.)

(8) If $f = 4$, or $f = 3$, choose x to make the last bracket on the right hand side of (6) equal to zero, and solve the corresponding pair of equations.

Extension You are familiar with the idea of a regular polygon – though you may not be used to thinking of the regular pentagram, with angles all equal to 36°, as a regular polygon. Investigate *semi*regular polygons'. What are the possibilities for the angles in a semiregular polygon? What can you say about the lengths of the sides?

5. (a) It is natural to imagine that, since squares and equilateral triangles can be used to construct regular tilings of the plane, cubes and regular tetrahedra can be used to construct regular tilings of space. We are all familiar with the way cubes fit together, face to face, to tile space. But what about regular tetrahedra?

(i) One way of deciding whether regular tetrahedra fit together without gaps to tile space is to picture what would happen round an edge in such a tiling. In the familiar tiling with cubes, four cubes fit together round each edge. This is possible because the angle between adjacent faces in each cube is exactly 90°, so four of them contribute exactly 360° round each edge. What is the angle between adjacent faces in a regular tetrahedron? Can a whole number of regular tetrahedra fit together without gaps round an edge?

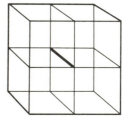

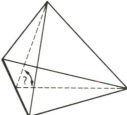

(ii) What is the angle between adjacent faces in a regular octahedron? A regular icosahedron? A regular dodecahedron?

(b) The angles you worked out in part (a) (i) and (ii) should enable you to work out which combinations of regular polyhedra can fit together without gaps round an edge. You can then test to see which of these combinations of regular polyhedra round an edge give rise to semiregular tilings of space.

Extension When we tile space with cubes we get exactly four cubes round each edge, so the 'edge figures' are all congruent. The other semiregular tiling you found in Question 5(b) also has its 'edge figures' all congruent. Suppose we want to tile space with *semi*regular polyhedra in such a way that the 'edge figures' are all congruent. There are exactly three other ways of doing this. What are they?

Investigation

(a)   Use the ideas in the Commentary on Question 2, bearing in mind that every face is an equilateral triangle. You will find it easier to check all possibilities systematically, using your hands as well as your brain, if you have access to plastic tiles like those produced under the trademark POLYDRON.

(b)   This is by no means easy. But you should be able to find some interesting infinite families of deltahedra.

30

Four examples

---

§1. Question 7. You may never have noticed, but it just so happens that

$$\sqrt{(2\tfrac{2}{3})} = 2\sqrt{(\tfrac{2}{3})}$$

How many other equations are there like this one?

SPEND AT LEAST FIFTEEN MINUTES ON THIS PROBLEM BEFORE READING ON. CLOSE THE BOOK!

Many will interpret the question as an invitation to guess. (How about $\sqrt{(3\tfrac{3}{4})} = 3\sqrt{(\tfrac{3}{4})}$?) But what if you did stumble on another equation of the same kind: what would that tell you? It might suggest some kind of pattern. But then again it might not! We need a more mathematical approach which is less dependent on lucky guesses.

Commentary There are lots. Try to find them all – and don't be afraid to use a little algebra.

It's not entirely clear what to do here. Should we try

$$\mathbf{A} \quad \sqrt{\left(n + \frac{n}{n+1}\right)} = n \times \sqrt{\left(\frac{n}{n+1}\right)} \ ?$$

$$Or \ \mathbf{B} \quad \sqrt{\left(m + \frac{n}{n+1}\right)} = m \times \sqrt{\left(\frac{n}{n+1}\right)} \ ?$$

$$Or\ \mathbf{C}\qquad \sqrt{\left(m + \frac{p}{q}\right)} = m \times \sqrt{\left(\frac{p}{q}\right)}\quad ?$$

$$Or\ \mathbf{D}\qquad \sqrt{\left(m + x\right)} = m \times \sqrt{x}\quad ?$$

A looks easiest – only one unknown! How can we get rid of those nasty square roots? Aha! Square both sides.

$$n + \frac{n}{n + 1} = n^2 \times \frac{n}{n + 1}$$

$$\frac{n + 2}{n + 1} = \frac{n^2}{n + 1}$$

so $\qquad n + 2 = n^2$

$$\therefore n = \underline{\quad\quad}\ ,\ or\ n = \underline{\quad\quad}$$

One of these is impossible. (Why?) So there aren't 'lots'! Why should we have expected this? (One quadratic equation and one unknown gives at most __ solutions.)

OK, let's try **B**. Squaring gives

$$m + \frac{n}{n + 1} = m^2 \times \frac{n}{n + 1}$$

$$m = (m^2 - 1) \times \frac{n}{n + 1}$$

> so $m \times (n + 1) = (m^2 - 1) \times n$
> Now $\mathrm{hcf}(m, m^2 - 1) = 1$ (Why?) and
> $\mathrm{hcf}(n, n + 1) = 1$,
> so $m = n$ and $m^2 - 1 = n + 1$.

$$\therefore m = n = \underline{\quad\quad}\ .$$

So there still aren't 'lots'. Perhaps you should have a go at **C**, making use of the clever idea in the box.
Then think about the most general interpretation of all – namely **D**.

§13. Question 1(a). Can you make six prime numbers which together use each of the nine digits 1–9 exactly once? How many different ways are there of doing this?

SPEND AT LEAST FIFTEEN MINUTES ON THIS PROBLEM BEFORE READING ON. CLOSE THE BOOK!

Commentary (1 is not a prime number.) Why can't 4 be the units digit of a prime number? How many of the digits 1–9 *could* be the units digit of a prime number? Write them down.

4, 6, 8 cannot be units digits. So the prime numbers must be __1, __2, __3, __5, __7, __9.

Which of these units digits cannot have a tens digit?

__2 would be a multiple of 2, and __5 would be a multiple of 5. So '2' and '5' must themselves be two of our prime numbers. We now have to complete the four endings

__1, __3, __7, __9

using just three extra digits 4, 6, 8 to make four other prime numbers. This means that there has to be at least one other single digit prime, so we want either

A: three single digit prime numbers, and three with two digits: i.e.

⟦2⟧, ⟦5⟧, ⟦3⟧, __1, __7, __9
⟦2⟧, ⟦5⟧, ⟦7⟧, __1, __3, __9

or

B: four single digit prime numbers, one with two digits and one with three digits i.e.

⟦2⟧, ⟦5⟧, ⟦3⟧, ⟦7⟧, __ __1, __9
or ⟦2⟧, ⟦5⟧, ⟦3⟧, ⟦7⟧, __ __9, __1

Now examine each case in turn. For example,
A: 49 and 69 are not prime numbers, so we must have

2, 5, 3, __1, __7, 89

giving just two solutions, or

2, 5, 7, __1, __3, 89

giving just one further solution.
Now look at case **B** in the same sort of way.

§16. **Question 4.** Is the large regular pentagon more than twice, or less than twice, the area of the star inside it?

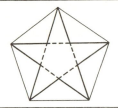

SPEND AT LEAST THIRTY MINUTES ON THIS PROBLEM BEFORE READING ON. CLOSE THE BOOK!

Commentary (i) You could try to do this by calculating each area approximately (using trig and a calculator) and then compare the two answers. But you should want to find a 'more mathematical' solution than that.

Let's just see what this means before looking for a 'more mathematical' solution. Here's one way.

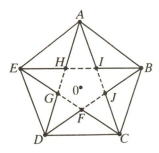

The angle at a vertex of a regular pentagon is ____°. So $A\hat{I}B = H\hat{I}J =$____°. $\triangle ABI$ is isosceles, so $B\hat{A}I =$ ____°. Clearly $A\hat{O}B =$ ____°, so $\triangle ABI$ has area $\frac{1}{2}.AB.(AB \tan 36)$, and $\triangle ABO$ has area $\frac{1}{2}.AB.(AB \tan 54)$. So by comparing tan 36 and tan 54 we should be able to decide whether the area of $\triangle ABO$ is more than, or less than, twice the area of $\triangle ABI$.

There are in fact several different ways of doing this; but a satisfying mathematical solution is harder to find.

It took me quite a while before I stumbled on the following. But a reasonable student should appreciate its superiority over mere calculation.

(ii)   One approach is to compare each of the five extra outside pieces, like ABI, with one fifth of the inside star;

What can you say about $\triangle FIB$? So how are the areas of $\triangle FIB$ and $\triangle AIB$ related? There are two natural ways of cutting up the inside star into five equal pieces. One of these produces a 'fifth' which can be compared very easily with $\triangle FIB$. Which is it? Once you answer this, the whole problem reduces to that of comparing the areas of the two triangles $\triangle IJO$ and $\triangle IJF$, which share a common base!

---

§12. Investigation 3.   How many odd numbers are there in the nth row of Pascal's triangle?

SPEND AT LEAST AN HOUR ON THIS – EXPERIMENTING, GUESSING, CHECKING, IMPROVING YOUR GUESSES, AND FINALLY TRYING TO PROVE WHAT YOU BELIEVE TO BE TRUE – BEFORE READING ON. CLOSE THE BOOK!

A brief discussion of an investigation such as this is bound to be more awkward than the previous three examples. My aim has been merely to show how considerable progress *can* be made provided one behaves reasonably intelligently. Readers who would like a more detailed and structured discussion of some specific mathematical investigations should refer to my book *Discovering Mathematics* (Oxford University Press, 1987).

The question posed is quite specific. Your task is to find some way of 'getting started' which holds out some hope of leading closer to your goal. The most obvious way of getting started is to write out the first ten or twelve rows of Pascal's triangle and see how many odd numbers there actually are in each row.

(a)   *Do this now.*

This may already have given you one or two ideas. But what next? In contrast to the problems which make up the bulk of this collection it is not at all clear what you should be trying to prove here. But while your ultimate goal may still be unknown, the evidence of the first few rows of Pascal's triangle should have drawn your attention to certain unexpected features. Now is the time to start trying to pull yourself up by your own bootstraps. See if you can say in advance how many odd numbers you expect to find in each of the next four rows.

(b) *Go on, have a go.*

Once you have made the best prediction you can manage (as opposed to an offhand guess which will cost you little if it turns out to be wrong), you should check to see how it stands up.

(c) *Well? How many odd numbers are there in each of the next four rows?*

So where should you go from here? Before proceeding, it may be worth taking a step back and trying to sort out what one should be trying to do in an investigation such as this.

• The original specific question led you first to explore what looked like sensible beginnings. ('How many odd numbers are there in each of the first so-many rows?')

• This gave rise to a number of seemingly pertinent observations. You were then in a position to risk yourself by making a few intelligent guesses. ('How many odd numbers do you expect to find in each of the next few rows? Which rows do you expect to have the most odd numbers?')

• You then tested these guesses against the facts. ('How many odd numbers are there really in each of the next few rows?')

• You can then start to define and improve those guesses which turn out to be interesting, but not quite right. ('OK, so the answer does seem to have something to do with powers of two. But what determines the power of two each time?')

• And when you finally think you understand what is going on, you must try to explain why it happens.

Without this final step the mathematical significance of the whole exercise is threatened. The precise extent to which individual students can successfully explain what they think they have discovered will, of course, vary from problem to problem. But they must be left in no doubt that, while intelligent guesswork is important, it does not constitute mathematics.

Unfortunately none of these generalities is going to help you to automatically improve on your own first tentative guesses in the present investigation. Indeed, there is no magic formula which can guarantee success. The most important thing is not to give up, but to keep on searching. However, there are a number of other things which sometimes help. One of these is the basic human urge to simplify and abbreviate the symbolism and notation being used, or to introduce a shorthand of one's own which seems particularly well suited to the problem. In this instance, many people realise sooner or later that the actual value of each entry in Pascal's triangle is irrelevant: all that matters is whether an entry is odd or even. This leads to a greatly simplified version of Pascal's triangle with only two sorts of entry. In particular, this makes it very easy to examine the first thirty-two, or even the first sixty-four rows.

(d) *Do this.*

It should then be possible to sort out *what* seems to be happening, and to predict accurately how many odd numbers you would expect to find in the sixty-fifth and sixty-sixth rows.

(e) *Well? How many do you expect there to be?*

(f) *And how many are there?*

(g) *Can you give a simple rule which predicts exactly how many odd numbers there will be in the nth row for any value of n?*

You must now try to explain why your rule works. (At this stage it becomes important to realise that there can be many different versions of the same rule, and that some versions are simpler than others. So be prepared to improve your own version, even if it seems to work!) Explaining why is rarely easy, but the following ideas should help.

(h) *Can you see how the odd/even pattern of the first two rows gets duplicated in the next two rows? Why does this happen?*

(i) *What do you notice in the next four rows? Explain.*

(j) *What is it about Pascal's triangle which gives rise to this 'self-replication'? Which rows appear to hold the key? (Look carefully at the first row, the second row, the fourth row, ...) How does this help to explain why your rule works?*

You should by now be beginning to wonder what is so special about odd and even. What do you think you would find if you looked at Pascal's triangle 'mod 3' instead of 'mod 2'?

31

Puzzling things out

Mainly for the teacher

This collection of problems was meant to convey a message. It should therefore have raised a number of questions about the nature of mathematical thinking, and about the way we learn and teach mathematics. This final section seeks to encourage the reader to reflect on some of these issues. In the process I shall make various remarks on the nature of school mathematics. This short final section is no place to attempt a complete and balanced analysis: it is a preliminary exploration of the relationship between ordinary school mathematics and the mathematics in this book. Many loose ends are left untied. It is, as it were, a Commentary on the pedagogical problems posed by the collection as a whole. And like the Commentary on the mathematical problems, it aims to give the reader 'something to think about' rather than to present a complete official solution.

The Spirit of Mathematical Puzzling
The contents of this book cannot be explored exhaustively in a single encounter. However, I shall assume that the reader, whether student or teacher, has already made a serious attempt at several of the sections.

Two main ideas run through the whole collection. The first idea is the distinction between:
    a one-step problem, and
    a multi-step problem.
Briefly stated, it is my contention that the latter kind of problem can generate rich mathematical thinking and activity, whilst the former kind tends to devalue both the pupil and the subject. Every teacher is confronted daily with a major challenge – namely to select multi-step problems which a given pupil, or class, can tackle successfully. The second idea, not entirely unrelated to the first, is that every learner should be continually encouraged to look for:
    connections and recurring themes.

That is, in Freudenthal's words, to experience mathematics as 'fraught with relations', rather than as consisting of so many unrelated rules, methods, and problems. Here the teacher's contribution is again crucial. It is not easy for a pupil to stand back while in the middle of tackling a problem, or group of problems, and to look for such connections. Yet the realization that new problems or techniques are often more (and sometimes less!) familiar than they seem is absolutely fundamental to mastering mathematics.

There is no foolproof way of distinguishing a one-step from a multi-step problem. The distinction is nonetheless important. To solve a one-step problem the solver need only implement one single step! In contrast, a multi-step problem requires the solver to identify one or more intermediate goals, or subproblems, *en route* to a solution. Thus, whether a particular problem is classified as a one-step or a multi-step problem will depend on the maturity and experience of the solver. For very young children the question 'What's eighteen and twelve?' may be a challenging multi-step problem, requiring the identification of various subgoals ('Ah! If I do eighteen and ten first . . . Mmm! . . . Twenty-eight! Then another two makes . . .'). But at some stage the procedures involved telescope into a single routine step. These routines certainly need to be practised – often using quick one-step questions. But such practice is only part of school mathematics. It should be a cardinal principle of all teaching and learning that: *as soon as a (type of) problem becomes a one-step problem for a particular solver, it is time to move on to something more demanding.*

The problems in this collection are aimed especially at those who regularly reach such a stage more quickly than most of their peers. However, *all* pupils, even when learning and practising standard techniques, need and deserve a richer mathematical experience than the endless sequence of routine one-step problems so often set. For example, using Pythagoras' theorem is not just a matter of finding the length of one side of a right-angled triangle given the lengths of the other two sides. One must also decide whether, and how, to apply it. Able pupils should as a matter of course meet problems in the spirit of those in Section 11, or Problem 2 in Section 22, or the following Problem A.

Problem A Find the area of this triangle.

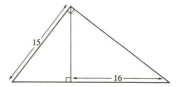

In the same vein, the notion of similarity and the elementary geometry of the circle should be exercised on questions like Problem 1 in Section 16, or the following Problems B and C.

Problem B   All three circles have radius 5 units. If *AD* is tangent to the third circle, find the length of *BC*.

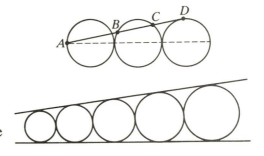

Problem C   Of the five circles shown, the smallest has radius 12 units and the largest has radius 18 units. Find the radius of the middle circle.

Before all this is dismissed as unrealistic, I hasten to admit that at present many able pupils find even the simplest multi-step problems too demanding – as numerous recent surveys have discovered. *But why is this?*

Where most pupils in a particular age-group fail to respond 'correctly' to a test question, it is fashionable to conclude that it is the mathematical content that is 'too hard'. Instead of simply reporting the observed flesh and blood responses, they are immediately translated into a supposed 'level of difficulty' for the mathematical content of the question. In most cases experience cries out against such a conclusion. A more obvious, and more disturbing, conclusion is that what children do with such problems reflects the way they have been *taught*. Problem D comes from one of the most influential surveys of recent years (whose very title – *Children's Understanding of Mathematics* – prejudges the source of the difficulties it uncovers).

Problem D   Area $= \frac{1}{3}$ cm²
Length $= ?$

Now one can imagine 15 year-olds having all sorts of difficulties here. But the actual percentage of correct responses can hardly be explained in terms of the difficulty of the mathematical content (as opposed to the way it is taught). The formula

area = length × breadth

for the area of a rectangle must be the best known in the whole of school mathematics, and the resulting equation

$$\frac{1}{3} = ? \times \frac{3}{5}$$

could be happily solved by many ten year-olds. In the event only one 15 year-old in twenty answered correctly.

The reasons in this particular case are unclear. What is clear is that the widespread practice of trying to get round the need for mathematical thinking by reducing everything as soon as possible to one-step routines such as: area = length × breadth, or the rule for dividing fractions, leads to a dependence on standard cues which is characteristically inflexible.

If pupils are to develop the confidence to puzzle out an approach to such problems for themselves, much work remains to be done. But it is neither the difficulty of the mathematics, nor the children's own 'understanding' which is holding us back.

Puzzles, Investigations, and 'Proper Mathematics'

The problems and exercises one uses to help pupils learn mathematics do not grow on trees. They have to be carefully chosen and adapted if they are to have the desired educational impact. Good mathematical problems are necessarily artificial. In contrast, 'realistic' problems tend to elicit 'realistic' responses involving little or no mathematics. In mathematics teaching, what matters is not whether a problem is plausibly real or artificial, but whether it is such that pupils are prepared to enter into the spirit of the mental world it conjures up. The problems in Section 25 depend entirely on this, as does the following.

Problem E   Hops, skips, and jumps are units of length. If b hops equal c skips, d skips equal e jumps, and f jumps equal g metres, how many hops are there in 1 metre?

This lovely question trips pleasantly off the tongue, and conjures up just the right sort of fanciful imagery. Fortunately most algebra problems do not have to be quite so striking. Once a pupil has entered into that curious world of algebraic manipulation and simplification, s/he can enjoy the starkest looking exercises provided only that they embody the right sort of challenge. The 'real world' of mathematics is a *mental* universe, and it is inhabited by *mental* objects and images.

Exactly the same criteria apply when one is choosing, or devising, problems for investigation.

Problem F   How many odd numbers are there in the nth row of Pascal's triangle?

Those who are sufficiently young at heart to enjoy the simple pleasure of generating Pascal's triangle for themselves have probably stopped reading and grabbed pencil and paper. (Why not have a go yourself before reading on?)

When choosing a problem for investigation by a group of pupils there are three crucial questions. Does the question evoke a sufficiently rich mental world to capture the pupils' potential interest? Will the rate at which curious features are likely to emerge be sufficient to maintain that interest? And is there enough accessible mathematics in the problem to make a satisfactory conclusion likely? The answers will depend on the pupils' own mental condition. (Are they prepared to explore *any* problem seriously on their own? Would they recognize a 'curious feature' if they saw one? What would they view as a 'satisfactory conclusion'?) A well-chosen problem should not only match pupils' present condition, but should stretch them in a way that will raise the level of their future performance.

There are, however, clear differences between the two kinds of mathematical activity – that is, between practising basic techniques and investigating questions like the one on Pascal's triangle. Even if one rejects the behaviourist approach to learning mathematics (in which pupils are trained to recognize and respond to standard cues in an endless sequence of routine one-step exercises), the problems one tackles when exercising basic techniques, and the spirit in which one approaches such problems, are relatively 'closed' and straightforward. In such problems one's final goal Z is usually specified, and the facts or techniques one needs to attain this goal should be clear from the instructional context. One's tasks is thus to choose an appropriate starting point A (if none is given) and to find a route from A to Z which involves a small number of steps of a very limited kind*. There is no reason to expect the solver to explore or extend such a problem in any way. The challenge consists in finding a starting point A and a route from A to Z. Once this has been done, it is time to try another problem.

A problem for investigation may be outwardly similar. It may, for example, appear to state a specific goal Z (as is the case with the Pascal's triangle problem). And, as before, we must first find a suitable starting point A, and proceed in some way until we eventually arrive at our goal Z. But there the similarity ends. There is no longer any real prospect of finding a quick and clever route to the specified goal Z. Thus in the absence of any obvious starting point A which might lead us to our goal in a logical, deductive way, we look for any kind of beginning, or point of entry, which might help us to understand the problem better. A deductive, mathematical solution will have to wait until we think we can see what exactly is going on. The specified goal Z still guides our progress, but in a rather different way.

Our attitudes as teachers to pupils' investigational work can easily be at variance with the attitudes we should be trying to encourage in ordinary classwork. In one sense one would welcome the suggestion that Problem A above 'could probably be solved by calling some length x and then using Pythagoras' theorem'. But one could scarcely be satisfied when an able pupil, having expressed such a view, either makes no attempt to implement the idea, or gives up too easily. Yet while we may realise that it is not enough to know how to solve Problem A 'in principle', it is all too easy to consistently reinforce bad habits by rewarding pupils' ideas and guesses prematurely. Our smiles and approving comments too often reveal our own sense of relief that a pupil, or group, is 'nearly there', when in fact an interested, but non-committal comment such as: 'Hmm! I suppose it might work. Have a go!' would be more appropriate.

---

*This may not sound very grand, but the mental coordination required even for mundane multi-step problems should not be underrated. Most of us need more, rather than less, of this kind of practice.

One has to learn to resist this temptation to reveal whether a pupil's guess is true or false on the basis of *one's own* pre-knowledge. The significance of such guesses lies in their relation to *the solver's own* problem solving process. Each such guess represents the solver's attempt to create order out of chaos by identifying some reasonable underlying pattern which seems to fit the available data. These guesses then provide a focus for subsequent exploration, as s/he seeks to disprove, refine, or confirm them, until at last they take the form of an apparently reliable assertion, which s/he can then try to prove. When we consistently use investigations for which the initial 'obvious' guess is correct, or when we betray by a smile or a frown (on the basis of our own pre-knowledge) the status of a pupil's crude first guess, it is as though we think we can short-circuit the whole painful process whereby truth has to be discovered. The challenge is to get students to look critically at their own guesses in the light of the evidence available *to them*. It is not the teacher's job to give their guesses some official seal – whether of approval or disapproval. When a guess is eventually obtained which seems to withstand all attempts to disprove it, pupils will be in a position to face up for themselves to the question of whether or not it is actually true. As long as the teacher's smile of approval remains the final arbiter, the question of 'proof' will never arise.

An understanding of the need for proof has to be learned. It does not come easily. It is the final link in an intricate network of ideas. Hunches are important, provided they are followed up. Guesses, once formulated, have to be tested against the facts. And what we believe to be true must be proved. Students who notice that the lengths 15 and 16 in Problem A suggest we are dealing with various multiples of a 3–4–5 right angled triangle have appreciated the value of hunches. If they then go on to check that lengths can be ascribed consistently to the unmarked sides on the basis of this guess, so much the better. But anyone who goes no further has failed to grasp the subtle interplay of intuition and deduction in mathematics. It is proof, or disproof, that has the final word.

A Gardiner

Answers

- These answers supplement the Commentary. Where an answer is incomplete, or omitted, consult the relevant Commentary.
- The answer to Question 3 in Section 7 is labelled **7.3**

QUESTIONS 1   **1.1**364   **2.1** e.g.(a)21 × 4 = 84   (b)14 × 3 = 42 (c)17 × 4 = 68
3.147   12   11   **4.1**63 × 63   **5.1**(a)169 × 7   (b)47 × 9   (c)64 × 6   **6.1**There is
none   **9.1**4   **10.1**e.g.(a)3 × 10 − 3 × 7 = 9   (b)2 × 8 − 2 × 5 = 6   4 × 5 −
2 × 8 = 4   2 × 8 − 3 × 5 = 1   3 × 5 − 1 × 8 = 7   (c)3 × 12 − 5 × 7 = 1
(d)5 × 8 − 3 × 13 = 1   **11.1**$\frac{1}{3}$ **13.1**(a)There are five solutions   (b)There are
thirteen solutions   **14.1**(a)5   (b)25   (c)125 (d)No (e.g.20)   No   **16.1**3:4
17.1(a)10   (b)(10 × 9)/2   (50 × 49)/2   **18.1**(a)1234567901234567901...
(b)37037037...   (c)Multiply by 3   **19.1**1 + 3n (n + 1)/2   **20.1**(a)Cube   (b)Opposite
edges equal   **21.1**(a)8   (b)9   16   (c)25   24   (d)$(2n − 1)^2$ 4 × 2n   **22.1**23   **23.1**(a)
Infinitely many   (b)Yes   Three   No   (c)No   None   **24.1**(a)91   161   171
(b)143   153   183   (c)117   147   187   (d)119   169   189   **25.1**Sunday   **26.1**(a)8
(b)40320 = 8!   **27.1**(a)Joins up   (b)Spirals away   Joins up   Joins up   Joins
up   Joins up   **28.1**There are 21   **29.1**(a)Two others.

QUESTIONS 2   **1.2**5   20;   24   **2.2**(a)21 + 81   88 + 22 or 82 + 28   79 + 24
or 74 + 29   (b)27 × 35   **3.2**4   6   65   8   26   12   21   6   28   8   11   16
4.28 = 4 + 4 and 18 = 6 + 6 + 6   **5.2**(a)Two solutions   (b)13 × 4   (c)54 × 3
(d)No   **6.2**5 and 7; it is conjectured (but not yet proved) that there are infinitely many
such pairs   **7.2**Most are easy   27 = (1 + 2) × ($\sqrt{3}$)^4 22 seems impossible
8.2 Only (f) can be done   **9.2**4   **10.2**(a)Get 1 (= 2 × 5 − 3 × 3), then add 3
(b)6 = 6 × 4 − 2 × 9   **11.2**Radius = 3 − 2 $\sqrt{2}$ **12.2**(b)84   9876   123456
82222228   **13.2**(a)Yes (excluding 00 01 04 09)   (b)Two   169   361   (c)One: 1681
(excluding 1600,2500 etc.)   (d)475   **14.2**(a)142857   No (e.g. 142857142857)
(b)285714   **15.2**(a)$2^{n+1} − 2^n = 2^n (2 − 1) = 2^n$ (b)$2^{n+1} − 2^{n-1} = 2^{n-1} (2^2 − 1) =$
$\frac{3}{2}× 2^n$ (c)$2^{n-1} × 2^{n+1} = 2^{2n} = (2^n)^2$ (d)Put x = 2 in $(x^{n-1} + x^{n-2} + . . . + x + 1)·$
$(x − 1) = x^n − 1$ **17.2**(a)15   (b)55 = 10 + 9 + 8 + . . . + 2 + 1
1275 = 50 + 49 + 48 + . . . + 2 + 1   **18.2**(a)13580246913580 . . .   4074074 . . .
(b)Multiply by 3   **19.2**(a)$761 = 20^2 + 19^2$ $200^2 + 199^2$ (b)25   9919   10587199
20.2(a)e.g. 1 24 120;   2 23 98;   3 22 78;   4 21 60;   5 20 44;   6 19 30
(b)e.g.   1 3 46 35;   1 3 97 24; 1 + a   3 − a   46 + a   35 − a
21.2(a)1   5 (b)24   32   **22.2**45/4   **23.2**(a)Square   (b)Yes   Infinitely many   No   No
(c)Yes Yes   (d)Yes   **24.2**2 457;   2 547   **25.2**84   **26.2**(a)14   (b)2^8
27.2(a)Joins up   (b)Spirals away   Joins up Joins up   Spirals away   Joins up
Spirals away   **28.2**(a)16   (b)41   (c)40.

QUESTIONS 3   **2.3**(a)25 × 3   (b)343 × 7   23 × 5 or 45 × 7   589 × 4
2572 × 4   3792 × 4   **3.3**None   **4.3**6   **5.3**(a)45)13085   (b)19)1045
(c)17)8619   (d)31)1023 or 33)1023   **7.3**All except 19   **8.3**8
9.30   1   1   2   2   2   3   3   3   4   4   4   4   **11.3**Radius = $\sqrt{2}$ − 1   **13.3**No
14.3(a)10344827586206896551724137931   (b)105263157894736842
210526315789473684   315789473684210526 etc.   **15.3**(a)$F_{n+2} − F_{n+1} = F_n$
(b)$F_{n+2} − F_n = F_{n+1}$ (c)$F_{n+1}F_{n-1} = F_n^2 − 1$ (why?)   (d)$F_{n-1} + F_{n-2} + . . . + F_1 =$
$F_{n+1} − 1$ (why?)   **16.3**(a)No   (b)Yes(why?)   **17.3**(a)First   (b)Neither
18.3 (a)24691358024691 . . .   740740 . . .   (b)Multiply by 3   **19.3**(a)30
(b)$22140 = 40^2 + 39^2 + . . . + 2^2 + 1^2$ **20.3**(a)50   6   25   38   (b)4   19   13   7;

26   1   47   35   No   No   No   **21.3**(a)15   (b)17   (c)10   (d)9   (e)12   (f)$8\frac{1}{2}$
(g)15   (h)1   (i)1   (j)10   (k)1   (l)1   **22.3**Cut along line joining centres of the two
rectangles   **23.3**(a)No   No   (b)Infinitely many   (c)Three   **25.3**5   35   60
26.3(a)8   (b)92   **28.3**(a)11   13;   10   11   (b)11   12;   10   11   (c)5   7   6   6;
3   5   7   6;   4   4   4   4   **29.3**10 other types (one with two different versions).

QUESTIONS 4   **1.4**$*=4$ **2.4**(a)689 × 5   2842 × 4   6753 × 3   **3.4**Two
colours:1(*c*)   2(*d, f, g, i*)   Three colours:2(*b, c, j*)   Four colours:1(*a,b*)   2(*a, e, h, k, l*)
4.4444444888889 = (666667)^2 **5.4**(a)12$\overline{)948}$ (b)12$\overline{)828}$ 13$\overline{)9217}$ 47$\overline{)10152}$
6.43   **7.4**All except 1,3,4,5   **8.4**(a)92836 + 12836   (b)3457 + 98636
96233 + 62513   9228 + 9228   850 + 850 + 29786   68782 + 68782 + 650
82526 + 19722 + 104   82524 + 19722 + 106   **9.4**(b)At least seven   At least six
10.4(a)Take G first   Then fetch *C*(or *W*) and take *G* back with you   (b) e.g. $\overleftrightarrow{C*CC}$,
$\overleftrightarrow{M*C*M}$, $\overleftrightarrow{M*C}$, $\overleftrightarrow{M*MM}$, $\overleftrightarrow{C*C}$ and its reverse   **11.4** Radius $= \frac{1}{3}(2\sqrt{(3)} - 3) \times$
circle radius $= (9 - 5\sqrt{(3)})/12 \times$ triangle side   **13.4** $\frac{1}{2}$
14.4 (a)112359550561797752809   (b)There is none
(c)366492146596858638743455497382   **15.4**(b)2^n (c)0 (except for first row)
(d)F_n **16.4**More (see Section 30)   **17.4**(a)15   (b)55   1275   **18.4**(a)11111...
(b)9999...   (c)9999...   (d)9999...   **19.4**(a)50   (b)105 = 7 × 6^2 × 5/12   563550
= 52 × 51^2 × 50/12   **20.4**(a)1   6   2   1   (b)15   18   1   12;   No;   6   4   5   7;
No   **21.4**(a) $\frac{1}{2}$ (b) $\frac{1}{2}$ (c) $\frac{1}{2}$ (d) $\frac{1}{2}$ (e) $\frac{1}{2}$ **22.4**Divide perimeter into five equal lengths
and join to centre   **25.4**28   21   **26.4**8   40320 = 8!   **29.4**5 regular and 13 other
types (three with two different versions).

QUESTIONS 5   **1.5**56 = 7 × 8   **4.5**10   **5.5**(a)47$\overline{)10387}$ (b)124$\overline{)10020316}$
7.5(a)21   (b)3^{21} (c)3^{421} (d)$8^{97654321}$ (e)No   **8.5**1777 + 1855 + 9234
1777 + 1855 + 9245   **10.5**(a)Three   (b)One (the middle one)   (c)Three
11.5$(2\sqrt{(2)}/\sqrt{(5 - \sqrt{(5)})} - 1) \times$ circle radius   **12.5**630   **14.5**Lots: 10125
101250   1012500 etc.   **15.5**(a)Every third   (b)Every fourth   (c)Every sixth   Every
fifth   Every twelfth   **17.5**(a)3 + 1   5 + 1   10 + 1   50 + 1   (b)6 + 1   15 + 1
1275 + 1   **18.5**(a)Recurring decimals   **19.5**(a)13   27   315 = $2^2 + 4^2 + 6^2 + 8^2 +$
$10^2 + 1 + 6 + 15 + 28 + 45 = 5$ (5 + 1) (4 × 5 + 1) / 2   4303 = $1^2 + 3^2 + 5^2 +$
... + 25^2 + 3 + 10 + 21 + ... + 300 = (12 + 1) (4 × 12^2 + 7 × 12 + 2)/2
(b)15   35 = 7!/4!3!   12!/4!8!   27!/4!23!   **20.5**(a)15 = (1 + 2 + 3 + ... + 9) / 3
(b)5(why?)   (c)10   **21.5**(b)(−2,1)   (−7,4)   (c)In line   Parallel   CF = FE = BA
(d)Same area   (e) $\frac{1}{2}$ $\frac{1}{2}$ **22.5**Nearly two hundred different ways
24.5(a)Multiples of 11 only   (b)Seven   **25.5**30   18   **26.5**8   Lots   **29.5**(a)1
(b)1 regular and 1 other.

QUESTIONS 6   **1.6**(a)1 and 6 top right and bottom left   (b)1 and 8 in the lower
two central circles   **4.6**52   **5.6**33   **6.6**2   **8.6**(a)2 + 97445 + 6928   (b)4$\overline{)14992}$
7$\overline{)67004}$ (c)5$\overline{)31770}$ 6$\overline{)18990}$ 6$\overline{)19002}$ 8$\overline{)35112}$ **9.6**(b)3   6   11   17   25   34
44   55   72   **11.6**Radius = 1 × circle radius = $((3 - \sqrt{(3)})/4) \times$ side length
12.6(a)Two: 1324   1432   (b)No   (c)Two: 2723625   2763675   (d)One: 36792
(e)No (multiple of 3 but not of 9)   (f)No (as for (e))   **13.6**(a)$N = 3 \times 6^3 = 2 \times 18^2$
(b)$N = 2^{15} \times 3^{10} \times 5^6$ **15.6**(a)They lie on two perpendicular lines
(b)$F_1^2 + F_2^2 + ... + F_{n-1}^2 = F_{n-1}F_n$(why?)   **17.6**(a)15   (b)35   120   2600   **20.6**All equal
21.6(b)(5,3)   (12,7)   (c)(7,4)   Equal   (d)Question 5 $\frac{1}{2}$ $\frac{1}{2}$ **22.6**Yes   8 **23.6**Half its
height   **24.6**(a)SOTS: 2   5   13   17   29   37   41   53   61   73   89   97
101   109   113   **25.6**The hour hand tells you how many minutes past the hour it is.
Then count back from the minute hand to find 12   **26.6**5.

QUESTIONS 7   **1.7**Lots (see Section 30)   **4.7**(a)Yes, 33856   (b)No   **5.7**(a)Two
solutions   (b)One solution   **6.7**5   Lots   **11.7**Radius = $(\sqrt{(3)} - 1) \times$ sphere
radius = $((\sqrt{(3)} - 1)/4) \times$ side length   **12.7**(b)One: 381654729   (c)One: 38165472

13.7(a)$\sqrt{(576)}$ = 24   No   (b)$\sqrt{(2916)}$ = 54   No   **15.7**(a)They approach the golden ratio $\tau = (1 + \sqrt{(5)})/2$ (b)They lie closer and closer to the line $x = \tau y$
18.7(a)Question 3(a)   (b)0.3333. . .   0.5555. . .   **20.7**(a)They go up (or down) in equal steps left to right   (b)All equal   **21.7**(a)In line   Same area $\frac{1}{2}$ $\frac{1}{2}$ (b)D = (7,8)
E = (7,7) $\frac{1}{2}$ (c)D = (6,8)   E = (7,7)   Yes   **22.7**(b)5 by 6   (c)8 by 8(why?)
25.7(a)22   (b)286 times a day.

QUESTIONS 8   **1.8**(a)18   Yes   (b)27   Yes   (c)36   48   **4.8** $\frac{1}{4}$ **6.8**7
11.8(a)Radius = $[\sqrt{(\frac{3}{2})} - 1]$ × sphere radius = $[\frac{2}{5} - \frac{3}{20} \sqrt{(6)}]$ × edge length
(b)Radius = $[\sqrt{(2)} - 1]$ × sphere radius = $[\sqrt{(2)} - 1]/[2 + \sqrt{(6)}]$ × edge length
12.8(a)99 or 00   (b)1089 or 0000   **13.8**(a)16   17   A suitable multiple of $2^3 \times 3^3 \times 5^2 \times 7 \times 11 \times 13 \times 19 \times 23 \times 29$ (b)4 or 5   **15.8**(a)$F_{n-2} \times a + F_{n-1} \times b$
(b)$c = a + b$ $d = c + b = a + 2b$ $e = d + c = 2a + 3b$ $f = e + d = 3a + 5b$
18.8(a)1   **20.8**(a)All equal   **21.8**(a)Equal and parallel   Equal areas   **25.8**No

QUESTIONS 9   **4.9**(a)4   (b)1 or 4   **12.9**(a)6   6   (b)Seven solutions, only one with missing digits equal   **13.9**Yes   **18.9**(a)377   610   **20.9**(a)5/3   6/7   7/3;
5/1   6/7   7/5;   5/6   6/3   7/4;   5/6   6/2   7/5;   5/3   6/2   7/8;   5/1   6/4   7/8   (b)Yes

QUESTIONS 10   **4.10**(a)300   **6.10**2   **13.10**One   42101000.

QUESTIONS 11   **4.11**(a)2   8   20   22   68   78   80   92   (b)$3^2 = 9$ **21.11**(b)44$\frac{1}{2}$

QUESTIONS 12   **4.12**(a)5   (b)Neither   **6.12**n prime   **21.12**$N^2 - 1$
$N = \sqrt{(a^2 + b^2)}$ with hcf(a,b) = 1.

QUESTIONS 13   **6.13**$n = 2^r$ **21.13**No ($\sqrt{(3)}$ is irrational).

QUESTIONS 14   **6.14**2.

QUESTIONS 16   **6.16**2.

QUESTIONS 17   **6.17**3.

QUESTIONS 19   **6.19**$n = 2^r$.

QUESTIONS 20   **6.20**Assume $n \geq 2$ (a)If $m^n - 1$ is prime, then $m = 2$ and n is prime   (b)If $m^n + 1$ is prime then $m = 1$, or m is even and $n = 2^s$.

Prime numbers
(less than 5000)

2	151	353	577	811	1049	1297	1559	1823
3	157	359	587	821	1051	1301	1567	1831
5	163	367	593	823	1061	1303	1571	1847
7	167	373	599	827	1063	1307	1579	1861
11	173	379	601	829	1069	1319	1583	1867
13	179	383	607	839	1087	1321	1597	1871
17	181	389	613	853	1091	1327	1601	1873
19	191	397	617	857	1093	1361	1607	1877
23	193	401	619	859	1097	1367	1609	1879
29	197	409	631	863	1103	1373	1613	1889
31	199	419	641	877	1109	1381	1619	1901
37	211	421	643	881	1117	1399	1621	1907
41	223	431	647	883	1123	1409	1627	1913
43	227	433	653	887	1129	1423	1637	1931
47	229	439	659	907	1151	1427	1657	1933
53	233	443	661	911	1153	1429	1663	1949
59	239	449	673	919	1163	1433	1667	1951
61	241	457	677	929	1171	1439	1669	1973
67	251	461	683	937	1181	1447	1693	1979
71	257	463	691	941	1187	1451	1697	1987
73	263	467	701	947	1193	1453	1699	1993
79	269	479	709	953	1201	1459	1709	1997
83	271	487	719	967	1213	1471	1721	1999
89	277	491	727	971	1217	1481	1723	2003
97	281	499	733	977	1223	1483	1733	2011
101	283	503	739	983	1229	1487	1741	2017
103	293	509	743	991	1231	1489	1747	2027
107	307	521	751	997	1237	1493	1753	2029
109	311	523	757	1009	1249	1499	1759	2039
113	313	541	761	1013	1259	1511	1777	2053
127	317	547	769	1019	1277	1523	1783	2063
131	331	557	773	1021	1279	1531	1787	2069
137	337	563	787	1031	1283	1543	1789	2081
139	347	569	797	1033	1289	1549	1801	2083
149	349	571	809	1039	1291	1553	1811	2087

2089	2393	2713	3041	3373	3697	4021	4363	4721
2099	2399	2719	3049	3389	3701	4027	4373	4723
2111	2411	2729	3061	3391	3709	4049	4391	4729
2113	2417	2731	3067	3407	3719	4051	4397	4733
2129	2423	2741	3079	3413	3727	4057	4409	4751
2131	2437	2749	3083	3433	3733	4073	4421	4759
2137	2441	2753	3089	3449	3739	4079	4423	4783
2141	2447	2767	3109	3457	3761	4091	4441	4787
2143	2459	2777	3119	3461	3767	4093	4447	4789
2153	2467	2789	3121	3463	3769	4099	4451	4793
2161	2473	2791	3137	3467	3779	4111	4457	4799
2179	2477	2797	3163	3469	3793	4127	4463	4801
2203	2503	2801	3167	3491	3797	4129	4481	4813
2207	2521	2803	3169	3499	3803	4133	4483	4817
2213	2531	2819	3181	3511	3821	4139	4493	4831
2221	2539	2833	3187	3517	3823	4153	4507	4861
2237	2543	2837	3191	3527	3833	4157	4513	4871
2239	2549	2843	3203	3529	3847	4159	4517	4877
2243	2551	2851	3209	3533	3851	4177	4519	4889
2251	2557	2857	3217	3539	3853	4201	4523	4903
2267	2579	2861	3221	3541	3863	4211	4547	4909
2269	2591	2879	3229	3547	3877	4217	4549	4919
2273	2593	2887	3251	3557	3881	4219	4561	4931
2281	2609	2897	3253	3559	3889	4229	4567	4933
2287	2617	2903	3257	3571	3907	4231	4583	4937
2293	2621	2909	3259	3581	3911	4241	4591	4943
2297	2633	2917	3271	3583	3917	4243	4597	4951
2309	2647	2927	3299	3593	3919	4253	4603	4957
2311	2657	2939	3301	3607	3923	4259	4621	4967
2333	2659	2953	3307	3613	3929	4261	4637	4969
2339	2663	2957	3313	3617	3931	4271	4639	4973
2341	2671	2963	3319	3623	3943	4273	4643	4987
2347	2677	2969	3323	3631	3947	4283	4649	4993
2351	2683	2971	3329	3637	3967	4289	4651	4999
2357	2687	2999	3331	3643	3989	4297	4657	
2371	2689	3001	3343	3659	4001	4327	4663	
2377	2693	3011	3347	3671	4003	4337	4673	
2381	2699	3019	3359	3673	4007	4339	4679	
2383	2707	3023	3361	3677	4013	4349	4691	
2389	2711	3037	3371	3691	4019	4357	4703	